AF815441

Biophysics and Diving Decompression Phenomenology

Authored By

B.R. Wienke

Los Alamos National Laboratory

Nuclear Weapons Technology Simulation and Computing

Program Manager Computational Physics

C & C Dive Team Ldr

Los Alamos, NM 87545

United State

Biophysics and Diving Decompression Phenomenology

Author: B.R. Wienke

ISBN (eBook): 978-1-68108-319-3

ISBN (Print): 978-1-68108-320-9 ©2016,

Bentham eBooks imprint.

First published in 2016.

advertisements or ideas contained in the Work.

Limitation of Liability:

In no event will Bentham Science Publishers, its staff, editors and/or authors, be liable for any damages, including, without limitation, special, incidental and/or consequential damages and/or damages for lost data and/or profits arising out of (whether directly or indirectly) the use or inability to use the Work. The entire liability of Bentham Science Publishers shall be limited to the amount actually paid by you for the Work.

General:

1. Any dispute or claim arising out of or in connection with this License Agreement or the Work (including non-contractual disputes or claims) will be governed by and construed in accordance with the laws of the U.A.E. as applied in the Emirate of Dubai. Each party agrees that the courts of the Emirate of Dubai shall have exclusive jurisdiction to settle any dispute or claim arising out of or in connection with this License Agreement or the Work (including non-contractual disputes or claims).
2. Your rights under this License Agreement will automatically terminate without notice and without the need for a court order if at any point you breach any terms of this License Agreement. In no event will any delay or failure by Bentham Science Publishers in enforcing your compliance with this License Agreement constitute a waiver of any of its rights.
3. You acknowledge that you have read this License Agreement, and agree to be bound by its terms and conditions. To the extent that any other terms and conditions presented on any website of Bentham Science Publishers conflict with, or are inconsistent with, the terms and conditions set out in this License Agreement, you acknowledge that the terms and conditions set out in this License Agreement shall prevail.

Bentham Science Publishers Ltd.
Executive Suite Y - 2
PO Box 7917, Saif Zone
Sharjah, U.A.E.
Email: subscriptions@benthamscience.org

BENTHAM SCIENCE

CONTENTS

FOREWORD

This book has a singular focus and intent and is a necessary addition to the library of any modeler, dive computer engineer, software designer, or table fabricator working in the diving arena. It is well written, concise, and probably the only full reference today on the extensive literature of applied decompression theory. Equations and models are complete and applications are keyed to the content of the book. The References are also extensive. The book follows up a number of earlier publications of the Author, and adds new material in the computational synthesis. Models, mathematical methods, statistical correlations, and wide ranging applications are the mainstay of presentation. Thumbs up all around on this timely and needed publication.

Tim O'Leary
NAUI Technical Diving Operations
Tampa
Florida
USA

PREFACE

This monograph covers a body of biophysics, gas transport, bubble studies and attendant models used for diving and hyperbaric applications, and divides into three Parts, namely; Biophysics And Models, Correlations And Validation, and Applications. Parts are all inter-connected by analysis. The biophysics of diving and decompression in the human body is extremely complex. More needs be learned to safely and routinely stage divers. The physics, biology, engineering, physiology, medicine, and chemistry of diving center on pressure, and pressure changes. The average individual is subject to atmospheric pressure swings of 3% at sea level, as much as 20% a mile in elevation, more at higher altitudes, and all usually over time spans of hours to days. Divers and their equipment can experience compressions and decompressions orders of magnitude greater, and within considerably shorter time scales. While effects of pressure change are readily quantified in physics, chemistry, and engineering applications, the physiology, medicine, and biology of pressure changes in living systems are much more complicated. Caution is needed in transposing biological principles from one pressure range to another. Incomplete knowledge and biophysical complexities often prevent extensions of even simple causal relationships in biological science. Gas exchange, bubble formation and elimination, and compression-decompression in blood and tissues in diving are governed by many factors, such as diffusion, perfusion, phase separation and equilibration, nucleation and cavitation, local fluid shifts, and combinations thereof. Owing to the complexity of biological systems, multiplicity of tissues and media, diversity of interfaces and boundary conditions, and plethora of bubble impacting physical and chemical mechanisms, it is difficult to solve the decompression problem *in vivo*. And equally difficult and elusive are direct measurements of bubbles, bubble sites, and effective transport properties of tissues and blood in living human systems. Early decompression studies adopted the medical supersaturation viewpoint. Closer looks at the physics of phase separation and bubbles in the mid-1970s, and insights into gas transfer mechanisms, culminated in extended kinetics and dissolved-free phase theories. In both cases, models are employed to stage divers as safely as possible to the surface. Optimally, these models ought be correlated with existing diving data and linked to the most current biophysics. So, the monogram describes underlying biophysics, connectivity to macroscopic models, and correlation with real diving data, with correlations as important as models. Applications to mixed gas, decompression, open circuit (OC), and rebreather (RB) diving are linked to a correlated bubble model for comparisons and risk analyses. Applications focus mainly on deep diving where risks increase and statistical collections and tabulations of data are very important.

B.R. Wienke
Los Alamos National Laboratory

Nuclear Weapons Technology Simulation And Computing
Program Manager Computational Physics
C & C Dive Team Ldr
Los Alamos
NM 87545
USA

ACKNOWLEDGEMENTS

It is a pleasure to thank my colleagues and friends over the past 35 years at Los Alamos National Laboratory, Livermore National Laboratory, the United States Navy, and Colleges and Universities in the United States and abroad for support. And it's also a pleasure to be affliate with Los Alamos National Laboratory in areas of particle and nuclear theory, particle transport, weapons physics and applications, Nevada Test Site underground testing and experiments, code development, computational physics, numerical methods, and high performance computing. Plus diving.

Special thanks to diving confidants, gurus, and researchers, namely Jim Joiner (Best Publishing), Dick Vann (Duke), Charlie Lehner (Wisconsin), Alf Brubakk (Trondheim), Brian Hills (Adelaide), Mike Powell (NASA), Bill Hamilton (Hamilton Research), Dave Yount (Hawaii), Tim O'Leary (NAUI Technical Diving), Alessandro Marroni (DAN), and Peter Bennett (UHMS).

CONFLICT OF INTEREST

The author confirms that author has no conflict of interest to declare for this publication.

Biophysics and Diving Decompression Phenomenology

2

CHAPTER 1

Biophysics and Models

Abstract: The biophysics of pressure changes in diving and decompression is extremely complex. Fundamental notions of underlying mechanisms of bubble formation and gas transport are presented. Models are discussed for staging diver ascents focusing on dissolved gas perfusion, bubble formation, and coupled mechanics. Oxygen toxicity is also quantified and detailed. The focus is computational models that have been encoded into underwater computers, tables, and dive planning software. There are some 11 models that have been implemented and safely used by divers, though not all have been tested nor validated.

Keywords: Bubbles, Dissolved gases, Models, Oxygen toxicity, Transport.

Gas exchange, bubble formation and elimination, and compression-decompression in blood and tissues are governed by many factors, such as diffusion, perfusion, phase separation and equilibration, nucleation and cavitation, local fluid shifts, and combinations thereof [17, 28, 31, 37, 44, 47, 48, 57, 61, 69, 87]. Owing to the complexity of biological systems, multiplicity of tissues and media, diversity of interfaces and boundary conditions, and plethora of bubble impacting physical and chemical mechanisms, it is difficult to solve the decompression problem *in vivo*. Early decompression studies adopted the supersaturation viewpoint. Closer looks at the physics of phase separation and bubbles in the mid-1970s, and insights into gas transfer mechanisms, culminated in extended kinetics and dissolved-free phase theories. Integration of both approaches can be useful. Phase and bubble models are more general than supersaturation models, incorporating their predictive capabilities as subsets. Indeed, for most recreational and nonstop diving, bubble and dissolved gas models collapse onto themselves, that is, they suggest similar staging regimens. Having said all that, data still plays the crucial role in model determination and

B.R. Wienke

applicability. Dive modeling is often more of an artform than science, and experiments directed at one or another aspect of unanswered diving questions can often produce divergent conclusions, further caveats, null results, and scattered-beyond-use data. Plus, macroscopic models cannot always cover all important aspects of microscopic phenomena. We cannot cover all model data correlations herein. Instead we indicate range of model use, sector use, history, and some sources for data correlation. We focus on just 11 models. There are more, but they do not enjoy the utility and range of applicability of the 11 discussed. This is not a critical review of models, as all models are incomplete today, but all herein enjoy varying degrees of success and utility. That is, these models with data modifications form the underpinnings of current diving tables, decompression computers, software, and associated protocols. The intent here is to present a working view of physical phase mechanics, then followed by a summary of working models in diving. Such discussion is neither medical nor physiological synthesis. Such aspects are omitted, and, for some, certainly oversimplified.

The physics, biology, engineering, physiology, medicine, and chemistry of diving center on pressure, and pressure changes. The average individual is subject to atmospheric pressure swings of 3% at sea level, as much as 20% a mile in elevation, more at higher altitudes, and all usually over time spans of hours to days. Divers and their equipment can experience compressions and decompressions orders of magnitude greater, and within considerably shorter time scales. While effects of pressure change are readily quantified in physics, chemistry, and engineering applications, the physiology, medicine, and biology of pressure changes in living systems are much more complicated. Caution is needed in transposing biological principles from one pressure range to another. Incomplete knowledge and biophysical complexities often prevent extensions of even simple causal relationships in biological science. With this, models of bubble formation in the body face a tough task.

For sake of connectivity to the medical-biological diving community. mixed (diving) units are employed. In such system pressure and depth are used interchangeably, that is, 1 *atm* = 33 *fsw*, with *fsw* denoting feet-of-seawater (pressure and depth).

CAVITATION AND NUCLEATION

Simply, *cavitation* is the process of vapor phase formation [5, 16, 29, 45, 58] of a liquid when pressure is reduced. A liquid cavitates when vapor bubbles are formed and observed to grow as consequence of pressure reduction. When the phase transition results from pressure change in hydrodynamic flow, a two phase stream consisting of vapor and liquid results, called a cavitating flow [3, 25, 63]. The addition of heat, or heat transfer in a fluid, may also produce cavitation nuclei in the process called boiling. From the physico-chemical perspective, cavitation by pressure reduction and cavitation by heat addition represent the same phenomena, vapor formation and bubble growth, usually in the presence of seed nuclei. Depending on the rate and magnitude of pressure reduction, a bubble may grow slowly or rapidly. A bubble that grows very rapidly (explosively) contains the vapor phase of the liquid mostly, because the diffusion time is too short for any significant increase in entrained gas volume. The process is called vaporous cavitation, and depends on evaporation of liquid into the bubble. A bubble may also grow more slowly by diffusion of gas into the nucleus, and contain mostly a gas component. In this case, the liquid degasses in what is called gaseous cavitation, the mode observed in the application of ultrasound signals to the liquid. For vaporous cavitation to occur, pressure drops below vapor pressure are requisite. For gaseous cavitation to occur, pressure drops may be less than, or greater than, vapor pressure, depending on nuclei size and degree of liquid saturation. In supersaturated ocean surfaces, for instance, vaporous cavitation occurs very nearly vapor pressure, while gaseous cavitation occurs above vapor pressure.

In gaseous cavitation processes, inception of growth in nuclei depends little on the duration of the pressure reduction, but the maximum size of the bubble produced does depend upon the time of pressure reduction. In most applications, the maximum size depends only slightly on the initial size of the seed nucleus. Under vaporous cavitation, the maximum size of the bubble produced is essentially independent of the dissolved gas content of the liquid. This obviously suggests different cavitation mechanisms for pressure (reduction) related bubble trauma in diving. Slowly developing bubble problems, such as limb bends many hours after exposure, might be linked to gaseous cavitation mechanisms, while rapid bubble problems, like central nervous system hits and and embolism immediately after surfacing, might link to vaporous cavitation. But it's certainly never been determined either way.

Now we know that the inception of cavitation in liquids involves the growth of submicroscopic nuclei containing vapor, gas, or both, which are present within the liquid, in crevices, on suspended matter or impurities, or on bounding layers [1, 9, 16, 18, 22, 29, 43, 56, 93]. The need for cavitating nuclei at vapor pressures is well established in the laboratory. There is some difficulty, however, in accounting for their presence and persistence. For a given difference between ambient and gas-vapor pressure, only one radius is stable. Changes in ambient, gas, or vapor pressures will cause the nuclei to either grow, or contract. But even if stable hydrostatically, bubbles and nuclei, because of constricting surface tension, will eventually collapse as gas and vapor diffuse out of the assembly. For instance, an air bubble of radius $10^{-3} cm$ will dissolve in saturated water in about 6 *sec*, and even faster if the water is under saturated or the bubble is smaller. In saturated solutions, bubbles will grow by diffusion, and then tend to be quickly lost at free surfaces as buoyant forces raise them up. A $10^{-2} cm$ air bubble rises at the rate of 1.5 *cm/sec* in water. If nuclei are to persist in water, or for that matter, any liquid media, some mechanism must prevent their dissolution or buoyant exit.

A number of possibilities have been suggested to account for the presence of persistent, or stabilized, nuclei in under saturated liquids, liquids that have been boiled, or denucleated. Crevices in the liquid, or surrounding boundary, may exert mechanical pressure on gas nuclei, holding them in place. Microscopic dust, or other impurities, on which gas and vapor are deposited, are stabilized already. Surface activated molecules, (such as hydrogen and hydroxyl ions in water), or surface activated skins formed from impurities may surround the nuclei and act as rigid spheres, offsetting constrictive surface tension, preventing diffusion of gas out of the nuclei and collapse. In all cases, the end result is a family, or group of families, of persistent nuclei. Time scales for stabilization and persistence of nuclei would obviously equate to the strength and persistence of stabilizing mechanism. Experimentally, trying to differentiate stabilization modes is difficult, because (eventual) growth patterns of nuclei are the same in all cases. The ultimate crumbling of surrounding shells, release of crevice mechanical pressure, removal of dust and impurity nucleation centers, and deactivation of surface chemicals can lead to the onset of bubble growth.

FLOW CAVITATION (REYNOLDS NUCLEATION)

Euler first studied cavitation in fluids, defining a cavitation number, κ. In a flowing fluid, the cavitation number, κ, is an indication of degree of cavitation, or tendency to cavitate [16, 59, 93]. Describing the similarity in the liquid-gas system, the cavitation number relates fluid pressure, p, to vapor pressure, p_v, through,

$$\kappa = 2\frac{p - p_v}{\rho u^2} \tag{1}$$

with ρ and u the fluid density and velocity. Cavitation and cavitating flows have long been of interest in shipbuilding and hydraulic machinery, underwater signal processing, propellor design, underwater detection, material damage, chemical processing, high pressure and temperature flows in nuclear reactors, volatility of rocket fuels, and bubble chambers for detection of high energy particles, to list a few. Cavitation processes in flowing blood and nearby tissue are of some interest to decompression modelers and table designers.

Any flow has a cavitation index, κ. If κ is very large, then p is sufficiently greater then p_v and/or u is fairly small. In such cases, the Reynolds number, Re, with a the cavitation void radius, u stream velocity, and η fluid viscosity,

$$Re = \frac{\rho u a}{2\eta} \tag{2}$$

is small and single phase flow is the result. However, if u is increased, and/or p is decreased, nucleation will occur at some value, κ_i, denoted the *incipient cavitation index*. Cavitation will then occur in the flowing fluid provided,

$$\kappa \leq \kappa_i \tag{3}$$

At ambient pressure, at 20 ^oC, water has vapor pressure of roughly 0.03 *atm*. At 100 ^oC, water has vapor pressure of 1.00 *atm* and boils. Experiments peg the onset cavitation velocity of water near 15 *m/sec*, at 20 ^oC, permitting direct estimation of κ_i, Plugging the above values into the cavitation index equation, we find for water,

$$\kappa_i = 0.873 \quad (4)$$

The density of water is a nominal 1.0 g/cm^3 above. If we consider water flowing at the same speed at 99 ^oC, with a vapor pressure of 0.99 *atm*, the cavitation index is roughly 100 times smaller, that is,

$$\kappa = 0.009 \quad (5)$$

At 100 ^oC, the cavitation index is obviously zero (spontaneous boiling) for all flow speeds.

Certainly, the cavitation index, κ, is a more complicated function of flow parameters, Reynolds numbers, vapor pressure, and boundary conditions, but the above approach is extensively employed in calibrating cavitating flows. As the cavitation index drops further and further below the incipient value, bubble production increases. In the above, the vapor pressure, p_v, is used, suggesting that the cavitation sites are filled with vapor. In reality, any combination of vapor and gas may fill the cavitation void, so the vapor pressure is replaced by p_c, the vapor-gas pressure,

$$\kappa = 2\frac{p - p_c}{\rho u^2} \quad (6)$$

The voids may indeed be filled with vapor, or backfilled with noncondensable wake gas just back of the cavitation site.

A number of notable cavitation patterns [16, 45, 58] breakout under flow cavitation and are of considerable interest to hydrodynamicist in high speed, high pressure flow regimes of propellors, airfoils and hydrofoils, turbine blades, pumps, reactor coolant flows, and fins, to name a few. Four generic cases (descriptors) are given below, but no connection with cavitation processes in the body is intimated nor linked therein:

- *cloud cavitation* (periodic nucleation) – is a frothy structure resulting from the interaction of a driving blade or fin with the primary flow wake, and is periodic likely due to shielding of vortices in the wake, or some other periodic fluctuation induced in the flow, or embodied within the flow;
- *sheet cavitation* (wake nucleation) – is a large scale cavitation carpet structure found on extended surfaces, resembles velour, and may be completely filled with vapor, and not necessarily a collection of individual bubbles;
- *foil cavitation* (lift nucleation) – is another extended region of cavitation vapor, or gas, sitting on the lifting surface (top) of an airfoil, or hydrofoil, reducing lift, and often forming vortex cavitation ring structures at the foil juncture of upper and lower surfaces;
- *vortex cavitation* (turbulence nucleation) – is a localized and focused, highly concentrated, vorticular cavitation bubble structure, often emanating from the sharp tips of propellors and fins in streamlines or rotating cascades, cause by pressure reductions within turbulent flow vortices.

Modeling cavitation processes is tedious and intensive numerically.

FRICTIONAL CAVITATION (TRIBONUCLEATION)

Tribonucleation has been demonstrated in the laboratory by Campbell and others [3, 20, 43, 65], that is, a process in which solid contacting plates, immersed in a fluid, are separated very rapidly with the formation of voids and cavitation seeds. Upon separation of hydrophobic and hydrophilic surfaces, a bridging vapor cavity has been observed. In a fluid, dimensionally, one expects hydrostatic tension, τ, to depend on fluid viscosity, η, the velocity of separation, u, the cross sectional area of contacting surfaces, A, and inversely on a small separation volume, V, that is,

$$\tau = \zeta \frac{\eta u A}{V} \tag{7}$$

with ζ a (data fit) constant. Viscous adhesion is another term describing negative tension, τ, developed when hydrophobic surfaces are pulled apart. On circular plates, experiments suggest, with R plate radius, h plate separation, and $\zeta = 1$,

$$A \propto 4\pi R^2 \quad \textbf{(8)}$$

$$V \propto \frac{4}{3}\pi h^3 \quad \textbf{(9)}$$

so that,

$$\tau = 3\eta u \frac{R^2}{h^3} \quad \textbf{(10)}$$

Linking adhesion to cavitation, one might expect voids when,

$$p - \tau \leq p - p_v \quad \textbf{(11)}$$

for p fluid gas tension, or simply,

$$\tau \leq p_v \quad \textbf{(12)}$$

Negative tensions, τ, are generated easily, For instance, in nominal water at 20 ^oC, plates roughly 1 *cm*, separated at 0.1 *mm*, and pulled apart with a separation velocity of 1 *cm/sec*, produce a negative tension of 0.03 *atm*, in the vicinity of water vapor pressure.

Evans and Walder [30] showed that denucleated shrimp were more prone to bubble formation upon decompression if muscular contractions were performed before decompression. Harvey [36] demonstrated that damaged excised tissue was more susceptible to bubble formation than healthy tissue. In plate experiments in olive oil-glycerol-water fluids, Ikels [43] generated tribonuclei between both hydrophobic and hydrophilic bilayers. This suggests that cavities *in vivo* might be formed by the rupture of cell membrane bilayers under exercise, as detailed by Powell [59]. Stress produced micronuclei in the body possibly occur with tissue surfaces coming together rapidly, and then withdrawing. Gas bubbles emerging from capillaries may have their genesis in the following way:

- muscle contractions compress the capillary walls;
- subsequent expansion of the capillary walls pushes back on muscle, changing dimensionality;
- separation of the wall endothelium produces a void momentarily, with an inward flux of active oxygen, water vapor, and carbon dioxide gases tending to stabilize the hole;
- when surrounding tissues are oversaturated with inert gas, the cavity experiences an indesign gradient, and grows in size;
- the initial hole is a (water) vaporous cavitation site, but later fills with gaseous components.

While such tribonucleation processes occur on time scales of seconds, the tribonuclei produced might last for hours, or so, depending on stabilization mechanisms offsetting constrictive surface tension. Also, the question of stabilization depends on the size of the tribonuclei, as very small ones might act like beebees and the large ones like soap bubbles under further muscle contraction-extension, gas loading, and tissue crevice shielding.

RADIATION CAVITATION (IONIZATION NUCLEATION)

The question of cosmic radiation is mentioned here in passing, as cosmic radiation is likely not a strongly contributing DCS factor. Cosmic rays (charged particles, ions) passing through matter ionize the surrounding media in the process of slowing and stopping. Large amounts of their kinetic energy are deposited locally in the ionization process. Cavitation voids can be formed as charge separated surfaces emerge. As energy deposition and ionization rates increase, so do cavitation processes. If the tissue surrounding the cavitation voids is supersaturated, the seeds so formed might grow. The same holds for radioactive elements present in the body, such as U_{238} which decays by nuclear fission and releases many charged particle fragments into tissue and blood. In their gel experiments, Yount and Kunkle [46, 68, 90 - 92] examined nucleation rates from cosmic rays, including neutrons, and found the collective contribution to be less than 0.01 nucleation sites per gel sample (roughly 2,5 *cm* square or round dishes), well below experimental sensitivities and counting rates. Fission rates in the body are also well below a fission per week.

CAVITATION HYSTERESIS (BUBBLE MEMORY)

If a set of cavitation nuclei, denoted by number, κ_i, are observed in a sample following any process reducing vapor pressure locally below ambient pressure, then restoration of ambient pressure above vapor pressure to extinguish (remove) all nuclei is not a reversible process [3, 16]. In fact, it's highly irreversible, in that, if the pressure is then restored to the initial value, the new set of cavitation nuclei, denoted κ_n, are different than the original distribution, κ_i. That is, $\kappa_n \neq \kappa_i$, akin to magnetic material hysteresis following a magnetization loop of materials to the same starting point. Cavitation processes do not conserve entropy, and hence, are highly irreversible. Cavitation materials retain a *cavitation memory*, likely due to structural changes following initial pressure reduction and cavitation void formation. The difference between the two quantities, κ_h, is called the *hysteresis index*,

$$\kappa_h = \kappa_i - \kappa_n \tag{13}$$

Cavitation hysteresis, without thinking about it much, might have implications for diving adaptation, reverse and repetitive profiles, enucleation, muscular movements, and effective perfusion rates. If, for any forward or reverse profile,

$$\kappa_n \leq \kappa_i \tag{14}$$

forward or reverse profiles might be indicated. While, if

$$\kappa_n \geq \kappa_i \tag{15}$$

forward or reverse profiles might be contraindicated. To date, some experiments [13, 91] *in vitro* suggest the former, that is,

$$\kappa_n \leq \kappa_i \tag{16}$$

for successively decreasing pressure titrations. Such suggests a reason for forward profile diving admonitions, apart from obvious reductions in gas loadings under the same. The question of reverse *versus* forward profiles (outside of recreational air diving above 130 *fsw*) still remains an issue for mixed gas and decompression diving. Cavitation hysteresis has neither been measured nor observed in tissue and blood.

HOMOGENEOUS AND HETEROGENEOUS NUCLEATION

In studying holes, or weaknesses, in liquid structures, two dominant cavitation mechanisms emerge [1, 4, 5, 13, 18, 29, 56, 68]:

- *homogeneous nucleation* – thermal motions within the liquid form temporary microscopic voids that produce nuclei rupturing the voids and subsequently growing into macroscopic bubbles;
- *heterogenous nucleation* – major weaknesses develop at boundary layers between liquid and solid container walls, or between liquid and small particles suspended in the liquid, facilitating rupture and bubble growth.

The seemingly simple dichotomy portends very complex processes in the real world, not easily measured nor quantified in bubble science, and especially not in living systems.

Homogeneous nucleation along established classical lines (within the kinetic theory of liquids) only permits one kind of weakness, namely, the transitory voids that develop due to the thermal motion of liquid molecules. In real systems, of course, several other types of weaknesses and dislocations are possible. Nucleation may occur at the junction of a liquid and solid body. Kinetic theories have been developed quantifying such heterogeneous processes, while also quantifying the relative probabilities for homogeneous *versus* heterogeneous nucleation at the same site. Heterogeneous nucleation may also occur on very small, submicron size contaminants in the fluid, with the size of the contaminants so small that differentiating homogeneous from heterogeneous nucleation is difficult to impossible in the laboratory.

Other weaknesses persist in liquids in the form of contaminant microbubbles, possibly present in crevices within boundaries, within suspended particles, or freely suspended in the liquid. These persistent weaknesses seem to resist dissolution completely. For instance, in water, microbubbles of air are virtually impossible to remove completely, perhaps due to interface contamination. While it's usually possible to remove such nuclei (denucleate) from small research samples, their presence persists in most engineering applications [1, 58], and likely biomedical systems [56]. Much focus has centered on water systems, but not other liquids in general.

Simple homogeneous nucleation theory works very well for certain liquids, such as ether and n-hexane, in which nearly perfect denucleation has been demonstrated. For water at room temperatures, the opposite has been the case [22, 26, 28, 33]. Homogeneous theory suggests a tensile strength for water near 1440 *atm*, corresponding to a critical radius of 10 *angstrom* ($10^{-3}\mu m$) for vapor phases formed by random motion of water molecules. The highest tensile strength observed for water is 277 *atm*, while the highest supersaturation pressure sustained without cavitation is 270 *atm* in helium gas. Both are about 5 times below theoretical predictions, and this is generally taken as evidence that impurities (*motes*) remain present in even highly denucleated media, like water. Experimental attempts to seed liquids with solid impurities have not been very successful. For example, Bateman and Lang [3] tried charcoal, ferric oxide, and sodium bicarbonate with inconclusive results. Fischer [32] found that salt attracts nuclei into solution, but then dissolves completely so that nuclei are likely gaseous, and not solid. Polystyrene spheres [22] in the fractional micron size range, thought to resemble motes in water, were unsuccessful in nucleating gelatin. Cloud seeding [50, 58] with iodine crystals has not produced rain in most circumstances. A consensus today is that smooth spheres of any size, hydrophobic or hydrophilic, only reduce the tensile strength of water by 25% at most.

Studies of the formation of vapor voids in pure liquids date back to the pioneering work of Gibbs, with modern twists provided by Bateman and Lang [3], and Yang [86] in the middle 1900s. The dynamics of homogeneous nucleation are fairly simple. In a pure liquid, surface tension is the intermolecular force holding molecules together, thus preventing the formation of large holes in the liquid. Liquid ambient pressure, P, exterior to the bubble surface, is lesser than interior bubble pressure, Φ, by an amount,

$$\Phi - P = \frac{2\gamma}{r} \quad \textbf{(17)}$$

where γ is the surface tension, and r is the bubble radius. The concept of surface tension (better yet, surface energy) has been shown to be a very accurate concept even down to a few intermolecular distances by Skripov and others [13, 65]. Perhaps such a simple description over a few molecular layers is surprising, but is nevertheless very useful.

The Gibbs free energy, G, for homogeneous nucleation processes is the sum of the energy deposited on the surface of the nucleus, $4\pi r^2\gamma$, plus the work done by the fluid to create the void, $4\pi r^3\ (p - p_b)/3$, with p_b the internal bubble pressure and p the external fluid pressure,

$$G = 4\pi r^2\gamma - \frac{4}{3}\pi r^3(p_b - p) \quad \textbf{(18)}$$

If a Laplace relationship is assumed across the bubble, we have,

$$p_b - p = \frac{2\gamma}{r} \quad \textbf{(19)}$$

and the usual expression for the free energy results. The work done by the fluid is not necessarily *ideal* dynamically, and other structures in, and around, the void may help or hinder the process. In such case, the Gibbs free energy is written,

$$G = 4\pi r^2\gamma - \frac{4}{3}\pi r^3\Psi \quad \textbf{(20)}$$

with Ψ a representation of the effective pressure difference across the bubble surface, $(p_b - p)$, for nonideal gas formation, external effects, bubble nonsphericity, and even heterogeneous impact on void formation. Some forms include,

$$\Psi = \beta(p_b - p) \qquad \textbf{(21)}$$

with β a pressure difference multiplier,

$$\Psi = (p_b - p) + \delta \qquad \textbf{(22)}$$

with δ an additive tissue compliance,

$$\Psi = \epsilon_b \ln \left[\frac{p_b}{p} \right] \qquad \textbf{(23)}$$

for change in volumetric free energy in an isothermal phase expansion, where α_b is the energy density in the bubble,

$$\Psi = \sum_{n=0}^{N} \alpha_n (p_b - p)^n \qquad \textbf{(24)}$$

a virial expansion of the pressure difference, which can be fitted to data like an equation-of-state.

Homogeneous nucleation processes occur in single component systems, while heterogeneous nucleation processes involve more than one component. To describe nucleation, a heterogeneous model, ascribed to Plesset [58], containing the homogeneous case as a subset, has been useful in applications. A solid hydrophobic sphere, of radius r_0, is surrounded by a concentric layer of vapor, out to a radius r. The instantaneous (Boltzmann) probability, dw, for the state depends on the difference in free energy, G, associated with the vapor phase,

$$dw = exp\ (-G/kT)\ dG \qquad \textbf{(25)}$$

at temperature, T, for (Gibbs) free energy change, G,

$$G = \frac{4}{3}\pi r^2 \gamma_{lv} + \frac{4}{3}\pi r_0^2 \left(\gamma_{vs} - \gamma_{ls}\right) \quad \textbf{(26)}$$

and γ_{lv}, γ_{vs}, and γ_{ls} surface tensions associated with the liquid-vapor, vapor-solid, and liquid-solid interfaces. The homogeneous case corresponds to $r_0 = 0$, that is, no solid and only liquid-vapor nucleation. This particular form of the Gibbs energy is extensively used in Monte Carlo simulations of bubble excitation [2, 21] and growth, homogeneous and heterogeneous.

Tensions, pulling parallel to their respective surfaces, at equilibrium have zero net component.

$$\gamma_{lv}\ cos\ \theta = \gamma_{vs} - \gamma_{ls} \quad \textbf{(27)}$$

with liquid-vapor contact angle, θ, measured through the liquid. Wetted (hydrophilic) solids exhibit acute contact angle, occurring when,

$$\gamma_{vs} - \gamma_{ls} > 0 \quad \textbf{(28)}$$

so that the meniscus of the liquid phase is concave. In this case, the solid has greater adhesion for the liquid than the liquid has cohesion for itself, the free energy required to maintain the vapor phase is large (because the solid surface tension term is positive), and the probability of nucleation is decreased by the solid impurity. For a nonwetting (hydrophobic) solid, the situation is reversed, that is, the contact angle is obtuse,

$$\gamma_{vs} - \gamma_{ls} < 0 \quad \textbf{(29)}$$

the meniscus is convex, the solid has less adhesion for the liquid than the liquid has cohesion for itself, the free energy is reduced because the solid surface tension term is negative, and the probability of formation is increased. In the limiting case, $cos\ \theta = -1$, the free energy is given by,

$$G = \frac{4}{3}\pi\gamma_{lv}\left(r^2 - r_0^2\right) \qquad \textbf{(30)}$$

which becomes small for cavity radius, r, near impurity radius, r_0.

In the above, a nucleation rate, j, is given by the Boltzmann energy partition function,

$$j = j_0 \; exp\left(-G_b\right) \qquad \textbf{(31)}$$

with G_b the Gibbs number,

$$G_b = \frac{G}{kT} \qquad \textbf{(32)}$$

for normalization factor over free energy, j_0,

$$j_0 = \rho\left[\frac{2\gamma}{\pi m^3}\right]^{1/2} \qquad \textbf{(33)}$$

with ρ the liquid density, and m the mass of the liquid molecule.

While theories of heterogeneous and homogeneous nucleation work well for a number of liquids, the application of the heterogeneous model to water with impurities is not able to reduce the tensile strength to observable values. Recall that the homogeneous theory of nucleation predicts a tensile strength of water near 1,400 *atm*, the heterogeneous theory, with a variety of solid impurities, drops the tensile strength down to 1,000 *atm*, and the measured value for water is approximately 270 *atm*. Yet, in any solution, gas nuclei can be deactivated (crushed) by application of large hydrostatic pressures. The process of *crushing* is also termed *denucleation*. When denucleated solutions are decompressed in supersaturated states, much higher degrees of supersaturation are requisite to induce bubble formation. In diving, denucleation has been suggested as a mechanism for acclimatization. If denucleation is size selective, that is, greater hydrostatic pressures crush smaller and smaller nuclei, and if number distributions of nuclei increase with decreasing radius (suggested by experiments), than a conservative deep dive, followed by sufficient surface interval, might in principle afford a margin of safety, by effectively crushing many nuclei and reducing the numbers of nuclei potentially excited into growth under compression-decompression. But this has not been proven in diving scenarios.

The mechanisms of nucleation in the body are obscure. Though nucleation most probably is the precursor to bubble growth, formation and persistence time scales, sites, and size distributions of nuclei remain open questions. Given the complexity and number of substances maintained in tissues and blood, heterogeneous nucleation would appear a probable mechanism. In that regard, the process of tribonucleation at tissue interfaces is a viable candidate for bubble seed production under even modest pressure reductions, certainly well below the 270 *atm* measured for water (watery tissue included).

MATERIAL RESPONSE

Under changes in ambient pressure, bubbles will grow or contract, both due to dissolved gas diffusion and Boyle's law. An *ideal* change under Boyle's law is symbolically written, denoting initial and final pressures and volumes with subscripts, *i* and *f*,

$$P_i V_i = P_f V_f \tag{34}$$

with bubble volume,

$$V = \frac{4}{3}\pi r^3 \tag{35}$$

for *r* the bubble radius. The above supposes totally flexible (almost ideal elastomers) bubble films or skins on the inside, certainly not unrealistic for thin skin bubbles. Similarly, if the response to small incremental pressure changes of the bubble skins is a smooth and slowly varying function, the above is also true in low order. Obviously, the relationship reduces to,

$$P_i r_i^3 = P_f r_f^3 \qquad \textbf{(36)}$$

for an ideal radial response to pressure change.

But for real structured, molecular membranes, capable of offsetting constrictive surface tension, the response to Boyle's law is modified, and can be cast in terms of Boyle modifiers, ξ,

$$P_i r_i^3 = P_f r_f^3 \qquad \textbf{(37)}$$

with ξ virial functions depending on *P*, *V*, and T . For thin and elastic bubble skins, $\xi = 1$. For all else, $\xi \neq$low order effect, or in the case that the bubble velocity (rising or sinking) is small. For gels studied in the laboratory, as an instance, surfactant stabilized micronuclei do not behave like ideal gas seeds with thin elastic films. Instead under compression- decompression, their behavior is always less than ideal. That is to say, volume changes under compression or decompression are always less than computed by Boyle's law, similar to the response of a wetsuit, sponge, tissue bed, or lung membrane. The growth or contraction of seeds according to an equation-of-state (EOS) is more complex than Boyle's law [24, 62], A *virial* expansion has for all *P*, *T*, *V* and mole fractions, *n*, for *R* the universal gas constant,

$$PV = nRT \sum_{i=0}^{N} \alpha_i \left[\frac{nT}{V}\right]^i \qquad \textbf{(38)}$$

or, treating the virial expansion as a Boyle modifier, ξ,

$$\xi PV = nRT \qquad \textbf{(39)}$$

across slowly varying data points and regions. Symbolically, the radius, *r*, can be cast,

$$r = \sum_{i=0}^{N} \beta_{i/3} \left[\frac{nRT}{P}\right]^{i/3} \tag{40}$$

or, again introducing Boyle modifiers, ζ,

$$\zeta r = \left[\frac{nRT}{P}\right]^{1/3} \tag{41}$$

for α and β standard virial constants. Obviously, the virial modifiers, ξ and ζ are the inverses of the virial sum expansions as power series. For small deviations from thin film bubble structures, both are close to one.

Observationally, though, the parameterization can take a different tack, following Yount [90]. In gel experiments, the EOS is replaced by two regions, the permeable (simple gas diffusion across the bubble interface) and impermeable (rather restricted gas diffusion across the bubble interface). In the permeable region, seeds act like thin film bubbles for gas transfer. In the impermeable region, seeds might be likened to beebees. An EOS of course recovers this response in both limits. Accordingly, just in gels, the corresponding change in critical radius, r, following compression, $(P - P_i)$, in the *permeable* region, satisfies a relationship,

$$(P - P_i) = 2(\gamma_c - \gamma) \left[\frac{1}{r} - \frac{1}{r_i}\right] \tag{42}$$

with γ_c maximum compressional strength of the surfactant skin, γ the surface tension, and r_i the critical radius at P_i. When P exceeds the structure breakpoint, P_c, an equation for the *impermeable* region must be used. For crushing pressure differential, $P - P_c$, the gel model requires,

$$P - P_c = 2(\gamma_c - \gamma) \left[\frac{1}{r} - \frac{1}{r_c}\right] + P_c + 2P_i + P_i \left[\frac{r_c}{r}\right]^3 \tag{43}$$

where,

$$r_c = \left[\frac{P_c - P_i}{2(\gamma_c - \gamma)} + \frac{1}{r_i}\right]^{-1} \quad \textbf{(44)}$$

is the radius of the critical nucleus at the onset of impermeability, obtained by replacing P and r with P_c and r_c above. Allowed tissue supersaturation, $\Delta\Pi$, is given by,

$$\Delta\Pi = 2 \;\frac{}{\gamma_c r}(\gamma_c - \gamma) \quad \textbf{(45)}$$

with, in the permeable region,

$$r = \left[\frac{(P - P_i)}{2(\gamma_c - \gamma)} + \frac{1}{r_i}\right]^{-1} \quad \textbf{(46)}$$

and, in the impermeable region,

$$r^3 - 2(\gamma_c - \gamma)r^2 - \frac{P_i}{\zeta} r_c^3 = 0 \quad \textbf{(47)}$$

for,

$$\zeta = P - 2P_c + 2P_i + \frac{2(\gamma_c - \gamma)}{r_c} \quad \textbf{(48)}$$

So, allowed supersaturation is a function of three parameters, γ, γ_c, and r_i. They can be fitted to exposures and lab data. But Boyle expansion or contraction needs be applied ad hoc to the excited seeds. Additionally, nuclei are assumed to generate over times scales, ω, such that,

$$r = r_0 + [1 - \exp\ (-\omega t)](r_i - r_0) \quad \textbf{(49)}$$

with r_0 the critical radius at initial time ($t = 0$). The fourth parameter, ω^{-1}, is on the order of minutes to hours, but was never really codified in the Yount gel experiments. In the blood and tissue of divers, regeneration of seeds links perhaps to exercise and activity, rather than to persistence over longer time scales, but the exact situation remains unclear [36, 68, 71].

BLOOD AND TISSUE BUBBLES

For watery blood (mostly water), densities are nominal,

$$1.00 \leq \rho \leq 1.15\ g/cm^3 \quad \textbf{(50)}$$

while surface tension depends upon lipid or aqueous preponderance in tissues and blood,

$$15 \leq \gamma \leq 80\ dyne/cm \quad \textbf{(51)}$$

with smaller values for lipids, and larger values for watery tissue. Similarly, the viscosity ranges,

$$0.0100 \leq \eta \leq 0.0400\ dyne\ sec/cm^2 \quad \textbf{(52)}$$

Considering blood flow speeds in the systemic circulation of the body, less than 25 *cm/sec* (vena cava), flow regimes for bubbles are mostly laminar, with low Reynolds numbers. Complicated patterns in and around heart valves are not laminar, but still low speed. Said another way, bubbles in blood flows can be treated mostly with Stokes parameterization (and simplicity).

BUBBLE FLOW REGIMES

Treating bubbles in the circulatory system as semi-solid spheres of radius *a*, we can estimate the onset of turbulent bubble flow as a function of bubble size and blood flow rate (varying considerably across the body proper). The smaller the bubble, of course, the higher the surface tension, and the solid sphere approximation gets better. Taking 0.50 as rough cutoff Reynolds number for laminar bubble flow in the body, we have for corresponding cutoff bubble radius, a,

$$Re = \frac{\rho u a}{2\eta} \leq 0.50 \quad \textbf{(53)}$$

so that,

$$a \leq \frac{2 \times 0.50\eta}{\rho u} \leq 0.0004\ cm \quad \textbf{(54)}$$

for nominal density and viscosity, and maximum blood flow. Roughly, bubbles smaller than 4 μm flow smoothly. At slower blood speeds, larger bubbles are accommodated, as well as at larger viscosities. Interestingly, intercellular cell separations range in the 10 - 20 μm range, likely permitting such flow regime. Flow speeds 5 times slower suggest laminar bubbles on the order of 20 μm here, impinging boundaries at the upper end of separations.

Bubbles are pliable and distensible, deviating from sphericity under increasing stress and flow pressure. Experiment and theory underscore gradual change from spherical bubbles to oblate bubbles as relative flow speeds between bubble and fluid media increase in magnitude. Beyond extreme bubble distortion, bubble fracture takes place. At opposite ends, in the one case, the surface tension and small bubble size make it relatively impervious to flow distortion, while in the latter case, the large bubble size accommodates considerable flow distortion.

BUBBLE FLOW FRACTURE

Similar to flow analysis, some ball park estimates on blood bubble breakup can be made. Surface tension, γ, of course varies across lipid and aqueous tissues, by some factor of 5 to 6, perhaps even more according to high pressure studies of thin film bubble permeabilities [2]. From the foregoing, we have the fracture cutoff, a, in watery flows,

$$a \geq \frac{4\gamma}{\omega\rho g u^2} \geq 0.0068 \ cm \tag{55}$$

and, in lipid flows,

$$a \geq 0.0408 \ cm \tag{56}$$

which are fairly large fracture radii. Below blood speeds of 25 *cm/sec*, the fracture radii increase considerably, obviously inversely as the flow speed squared. It would thus appear from these back of the envelope estimates that flow fracture is unlikely for 100 μm bubbles, and smaller.

BUBBLE SHOCK FRACTURE

Rapidly applying high pressure to a liquid will fracture bubbles and seeds. Tapping a pressurized container of liquid will often eliminate bubbles (denucleation) when pressure is released. This is often done with soft drinks. Such procedures generate pressure waves in the liquid which can fracture seeds. Treating the bubbles as an ideal gas, and denoting the pressure differential across the shock wave, $P_f - P_i$, with f and i denoting final and initial shock pressures, we have from the Rankine-Hugoniot jump equations [25], with c_P and c_V specific heats at constant pressure, P, and constant volume, V, M the Mach number (shock speed divided by sound speed in the media), and $M > 1$,

$$P_f - P_i = P_i \left[\frac{2c_P/c_V}{c_P/c_V + 1} (M^2 - 1) \right] \geq \frac{2\gamma}{r} \tag{57}$$

as a simple mechanical estimate for a pressure wave to exceed bubble surface tension. Taking $c_P / c_V = 5/3$ for an ideal gas, we see,

$$P_i \geq \left[\frac{8\gamma}{5r(M^2 - 1)} \right] \tag{58}$$

yielding a threshold pressure, P_i, for fracturing a bubble of radius, r. Turning the equation around, the threshold radius, r, for fracture under shock passage with initial pressure, P_i, is given by,

$$r \geq \left[\frac{8\gamma}{5P_i(M^2 - 1)} \right] \quad \textbf{(59)}$$

If the ratio of shock speed to sound speed, M, is small, then large pressures, P_i, are necessary for fracture. Or, only large bubbles are fractured when the Mach number, M, is small.

BUBBLE BUOYANCY

Gas bubbles in fluids will usually float, but very small bubbles with high density gas trapped inside can sink. In blood and tissues, a cutoff radius for floating and sinking bubbles can be estimated in an approximate sense, treating viscosity as a low order effect, or in the case that the bubble velocity (rising or sinking) is small. For watery tissue and blood, with nominal density, $\rho = 1.15\ g/cm^3$, bubbles of density ρ' will rise provided,

$$\rho' \leq 1.15\ g/cm^3 \quad \textbf{(60)}$$

or, denoting the mass of trapped gas in the bubble, m',

$$m' < \frac{4 \times 1.15}{3} \pi a^3\ g = 4.82 a^3\ g \quad \textbf{(61)}$$

A bubble will sink, alternatively, for the opposite case,

$$m' > \frac{4 \times 1.15}{3} \pi a^3\ g = 4.82 a^3\ g \quad \textbf{(62)}$$

Bubble masses in the body are thought to roughly range,

$$10 \leq m' \leq 0.10 \times 10^{-9}\ g \quad \textbf{(63)}$$

Most bubbles in physical systems tend to float excepting very dense microemulsions possessing more of a solid structure than gaseous makeup [16,

22, 90]. These are sometimes called $anti-bubbles$.

Putting the above together, a few comments about body bubbles seem clear. Bubbles with radii larger than listed, for given mass, will rise, while those with radii smaller than listed, will sink in blood and watery tissue. Recall that 1 μm = $10^{-4} cm$, so bubbles in the 1 to 15 *micron* range exhibit enormous surface tension in the body, with some floating up to the capillary boundaries and some sinking. Some will tumble along circulatory boundaries, while others will be be destroyed by the contact. Others, near hydrodynamic and static equilibrium, will be carried along by the flow. New bubbles, created by cavitation, nucleation, and crepitation, may dump into the blood flow, or remain trapped at birth sites. Large bubbles in the blood flow may fracture spontaneously, or possibly be eliminated by the filter systems of the lung and pulmonary circulation. Some (very small) may survive all destruction mechanisms, and pass back out into the arterial and systemic circulation. The scenarios are complex, and virtually intractable. As pressure changes, bubbles support gradients for gas transport across thin film boundaries, changing bubble sizes and properties along the way. Even without pressure changes, constrictive surface tension will tend to collapse bubbles with radii less than critical values.

BLOOD AND TISSUE CAVITATION

As seen for water, the incipient cavitation index, κ_i, is of order of 1.0, at room temperature and sea level pressure (1 *atm*). This offers a good baseline for spontaneous cavitation estimates in flowing blood, laminar and turbulent. Taking a maximum blood flow rate of 30 *cm/sec*, and treating blood as water (density), the cavitation index at sea level can be first estimated. Taking water vapor pressure at body temperature to be as high as 0.10 *atm*, we see easily,

$$\kappa = 2024.50 \quad \textbf{(64)}$$

while at 18,000 *ft* elevation, with pressure roughly 0.5 *atm*, we find,

$$\kappa = 900.66 \quad \textbf{(65)}$$

Both are far above the incipient value, and laminar blood flow cavitation is highly unlikely. Blood flows with a rotational component, possibly with small vortices in and around heart valves and bending constrictions, support higher cavitation numbers, but not below the incipient index. If vortex pressure reduction is even 25% of ambient pressure, all other factors the same, the cavitation index is roughly at sea level,

$$\kappa = 337.75 \quad \textbf{(66)}$$

and at 18,000 *ft* elevation,

$$\kappa = 56.29 \quad \textbf{(67)}$$

Both cavitation numbers are still far above the incipient value, and rotational blood flow cavitation remains highly unlikely. Vortex pressure would need be reduced to roughly 0.1005 *atm* to drop the cavitation index near unity, and below. Suffice to say that flow cavitation in the blood requires ambient pressure reductions near vapor pressure for the very low blood speeds in the body. Even if blood densities are doubled, cavitation numbers are only halved, and cavitation in highly lipid blood flows remains unlikely. Tissue cavitation is another question, with a more likely probability.

In tissues, cavitation seems likely as tissue surfaces rub together. Tribonucleation might manifest itself in a number of sites in biological systems. Numerous sites exist where surfaces may contact each other and rub across one another. Articulating joints are a possibility, whereby synovial fluid may display a high enough viscosity to mimic experiments in the laboratory, as suggested by Ikels. Small circulatory vessels may collapse and expand in response to muscle movement, as well as ligatures and tendons connected to bone joints. Negative tensions, τ, as low as 0.10 *atm* support bubble formation between plates submersed in oil and separated with velocity as low as 0.7 *cm/sec*, The higher the separation speed, the greater the number of bubbles formed in experiments. We can use these results to estimate muscle separation speed, u, or contact ratio, χ, that is, the ratio of muscle surface area, A, to separation volume, V,

$$\chi = \frac{A}{V} \quad \textbf{(68)}$$

for tribonucleation in body tissue interfaces. First, we have,

$$\tau \leq \eta u \chi \tag{69}$$

or,

$$u \geq \frac{\tau}{\eta \chi} \tag{70}$$

Taking tissue (fluid) viscosity, $\eta = 0.02$ *dyne sec/cm*^{2}, a value between water and glycerol, $\chi = 2 \times 10^7 cm^{-1}$ from experiments by Campbell [20], and $\tau = 0.10$ *atm*, we estimate (minimal) muscle separation speed, u,

$$u \geq 0.25 \ cm/sec \tag{71}$$

This targets a speed range of muscle motion with small velocity for producing tribonuclei according to adhesion dynamics. Rubbing tissues are certainly capable of generating negative tensions in the *atm* range, as suggested in kinesiological studies [43]. The contact ratio, χ, is a crucial controlling factor in tribonucleation. It can vary rapidly for different surface separations in fluids, that is, small changes in surface separations induce large changes in the ratio. Obviously, χ is unknown for tribonucleation in the body, though η and u can be reasonably estimated.

If the fluid media supports gas tension, p, bubble formation is enhanced, and a Laplacian relationship is assumed,

$$p - \tau = \frac{2\gamma}{r} \tag{72}$$

with the tension reduced by the negative void pressure, τ . Bubble sizes vary inversely as the pressure differences, as verified by many investigators [1, 4, 5, 20, 50, 91].

MICRONUCLEI DISTRIBUTIONS AND LIFETIMES

Most questions of seed distributions, lifetimes, persistence, and origins in the body are unanswered today [18, 30, 36, 39, 41, 55, 69, 77, 78]. And while we have yet to measure microbubble distributions and lifetimes in the body, we can gain some insight from laboratory measurements and statistical mechanics. Microbubbles typically exhibit size distributions that decrease exponentially in radius, *r*. Holography measurements of cavitation nuclei in water tunnels suggest, fitting data from Mulhearn [50],

$$N = N_0 \ exp \ (-\beta r) \tag{73}$$

with,

$$N_0 = 1.017 \times 10^{12} \ m^{-3} \tag{74}$$

$$\beta = 0.0512 \ \mu m^{-1} \tag{75}$$

The Yount experiments in gels also display exponential dependences in cavitation radii,

$$N = N_0 \ exp \ (-r/\alpha) \tag{76}$$

with,

$$N_0 = 662.5 \ ml^{-1} \tag{77}$$

$$\alpha = 0.0237 \ \mu m \tag{78}$$

Both MRI and Doppler laser measurements of water and ice droplets [45] in the atmosphere underline exponential decrease in number density as droplet diameter increases. Ice and water droplets in clouds typically range, 2 $\mu m \le r \le 100 \ \mu m$. Dust and pollutants [51] are also exponentially distributed, potentially serving as heterogeneous nucleation sites. It might be a surprise if micronuclei in the body were not exponentially distributed in number density *versus* size.

The lifetimes of cavitation voids are not known, nor measured, in the body. The radial growth equations provide a framework for estimation using nominal blood and tissue constants. Consider first the mass transfer equation,

$$\frac{\partial r}{\partial t} = \frac{DS}{r}\left[\Pi - P - \frac{2\gamma}{r}\right] \qquad \textbf{(79)}$$

with all quantities as before, that is, r bubble radius, D diffusivity, S solubility, γ surface tension, P ambient pressure, and Π total gas tension. The time to collapse, τ, can be obtained by integrating over time and radius, taking initial bubble radius, r_i,

$$\tau = \int_0^\tau dt = \int_{r_i}^0 \left[\frac{r}{DS}\right]\left[\frac{1}{\Pi - P - 2\gamma/r}\right] dr = \left[\frac{\Delta p r_i(4\gamma + \Delta p r_i) - 8\gamma^2 \ln\ (1 - \Delta p r_i/2\gamma)}{2DS\Delta p^3}\right] \qquad \textbf{(80)}$$

with,

$$\Delta p = P - \Pi \qquad \textbf{(81)}$$

If surface tension is suppressed, we get,

$$\tau = \frac{r_i^2}{2DS\Delta p} \qquad \textbf{(82)}$$

In both cases, small tension gradients, Δp, and small transport coefficients, DS, lead to long collapse times, and vice-versa. Large bubbles take a longer time to dissolve than small bubbles. Taking nominal transport coefficient for nitrogen, $DS = 56.9 \times 10^{-6} \mu m^2/sec\ fsw$, and initial bubble radius, $r_i = 10.0\ \mu m$, for $\Delta p = 3.0\ fsw$, and $\gamma = 40\ dynes/cm$, we find,

$$\tau = 0.25\ sec \qquad \textbf{(83)}$$

In the Rayleigh-Plesset picture, the radial growth equation takes the form, neglecting viscosity,

$$\left[\frac{\partial r}{\partial t}\right]^2 = \frac{2(\Pi - P)}{3\rho}\left[\frac{r_i^3}{r^3} - 1\right] + \frac{2\gamma}{\rho r}\left[\frac{r_i^2}{r^2} - 1\right] \qquad \textbf{(84)}$$

so that the collapse time by diffusion only is,

$$\tau = \int_0^\tau dt = \left[\frac{3\rho}{2(\Pi - P)}\right]^{1/2} \int_{r_i}^0 \left[\frac{r_i^3}{r^3} - 1\right]^{-1/2} dr = r_i \frac{\Gamma(5/6)}{\Gamma(1/3)} \left[\frac{3\pi\rho}{2\Delta p}\right]^{1/2} \quad (85)$$

with,

$$\Gamma(5/6) = 1.128 \quad (86)$$

$$\Gamma(1/3) = 2.679 \quad (87)$$

Suppressing the diffusion term in the growth equation, there similarly obtains,

$$\tau = \int_0^\tau dt = \left[\frac{\rho}{2\gamma}\right]^{1/2} \int_{r_i}^0 r^{1/2} \left[\frac{r_i^2}{r^2} - 1\right]^{-1/2} dr = r_i \frac{\Gamma(-3/4)}{\Gamma(-1/4)} \left[\frac{\pi\rho r_i}{4\gamma}\right]^{1/2} \quad (88)$$

with,

$$\Gamma(-3/4) = -4.834 \quad (89)$$

$$\Gamma(-1/4) = -4.062 \quad (90)$$

Collapse time in the Rayleigh-Plesset picture is linear in initial bubble radius, r_i, and inversely proportional to the square root of the tension gradient, Δp, or the surface tension, γ. Taking all quantities as previously, with density, $\rho = 1.15$ g/cm^3, we find with surface tension suppressed,

$$\tau = 2.91 \times 10^{-3} \ sec \quad (91)$$

and, for the diffusion term term suppressed with only the surface tension term contributing,

$$\tau = 2.52 \times 10^{-6} \ sec \quad (92)$$

Dissolution times above range,

$$10^{-6} \ sec \leq \tau \leq 10^{-1} \ sec \quad (93)$$

In the Yount model of persistent nuclei, within the permeable gas transfer region, seed nuclei lifetimes, τ, range,

$$10^{-6}\ sec \leq \tau \leq 10^{-2}\ sec \tag{94}$$

The collapse rate increases with both γ and Δp, and inversely with r_i. Small bubbles collapse more rapidly than large bubbles, with large bubble collapse driven most by outgassing diffusion gradients and small bubble collapse driven most by constrictive surface tension. Between these extrema, both diffusion and surface tension play a role. In any media, if stabilizing material attaches to micronuclei, the effective surface tension can be reduced considerably, and bubble collapse arrested temporarily, that is, as $\gamma \rightarrow 0$ as a limit point. For small bubbles, this seems more plausible than for large bubbles because smaller amounts of material need adhere. For large bubbles, bubble collapse is not aided by surface tension as much as for small bubbles, with outgassing gradients taking longer to dissolve large bubbles than small ones. In both cases, collapse times are likely to lengthen over the short times estimated above. Additionally, external influences on the bubble, like crevices and surface discontinuities, may prevent bubble growth or collapse. All this adds to the bubble complexities faced by hydrodynamicists and modelers.

COMPUTER ALGORITHMS

Decompression science and application to diving is an ongoing effort. The systematics of gas exchange [11, 14, 39, 45, 59, 79], nucleation [3, 4, 14, 29, 31, 42, 71], bubble growth [7, 11, 28, 66, 71, 90] and elimination [22, 26, 50, 54, 55], counter diffusion [40, 46, 82, 89], oxygen impact [34, 51, 52, 72], and adaptation [12, 13, 30, 36, 41] upon diving decompression staging [15, 30, 39, 44, 57, 64], and attendant altitude modifications [5, 16, 27, 39, 70, 77] are so complicated that theories only reflect pieces of the puzzle. Computational algorithms, tables, and manned testing are, however, requisite across a spectrum of activities. And the potential of electronic devices to process tables of information or detailed equations underwater is near maturity, with virtually any algorithm amenable to digital implementation. Pressures for even more sophisticated algorithms are expected to grow.

Still computational models enjoy varying degrees of success or failure. More complex models address a greater number of issues, but are harder to codify in decompression tables. Simpler models are easier to codify, but are less comprehensive. Some models are based on first principles, but most are not. Application of models can be subjective in the absence of definitive data, the acquisition of which is tedious, sometimes controversial, and often ambiguous. If deterministic models are abandoned, statistical analysis can address the variability of outcome inherent to random occurrences, but mostly in manner indifferent to specification of controlling mechanisms. The so called dose-reponse characteristics of statistical analysis are very attractive in the formulation of risk tables. Applied to decompression sickness incidence, tables of comparative risk offer a means of weighing contributing factors and exposure alternatives.

Bubble models address the coupled issues of gas uptake and elimination, bubbles, and pressure changes in different computational approaches. Application of a computational model to staging divers and aviators is often called a diving algorithm. Consider the computational model and staging regimen for some published algorithms, namely, diffusion, perfusion, thermodynamic, varying permeability, reduced gradient bubble, tissue bubble diffusion, and linear-exponential phase algorithms. Dissolved gas models are listed first, followed by dual phase and bubble models thereafter.

Dissolved gas diving algorithms historically trace back to the original Haldane experiments in the early 1900s. They are still around today, in tables, meters, and diving software. That is changing, however, as modern divers go deeper, stay longer, decompress, and use mixed gases.

MULTITISSUE MODEL (MTM)

The multi-tissue model (MTM) was originally proposed by Haldane [15], with modern correlations and tunings [6, 12, 37, 44, 61, 86] dating back to the 1950s. Exchange of inert gas, controlled by blood flow rates across regions of varying concentration, is driven by the gas gradient, that is, the difference between the arterial blood tension, p_a, and the instantaneous tissue tension, p. This behavior is modeled in time, t, by classes of exponential response functions, bounded by p_a

and the initial value of p, denoted p_i. These multitissue functions satisfy a differential perfusion rate equation,

$$\frac{\partial p}{\partial t} = -\lambda(p - p_a) \tag{95}$$

and take the form, tracking both dissolved gas buildup and elimination symmetrically,

$$p - p_a = (p_i - p_a)\ exp\ (-\lambda\ t) \tag{96}$$

$$\lambda = \frac{0.693}{\tau} \tag{97}$$

with perfusion constant, λ, linked to tissue halftime, τ. Compartments with 1, 2.5, 5, 10, 20, 40, 80, 120, 180, 240, 360, 480, and 720 minute halftimes, τ, are employed, and halftimes are independent of pressure. Compartments correlate roughly from brain to bone, that is, fast (few minute halftimes) to slow (few hour halftimes) tissues.

In a series of dives or multiple stages, p_i and p_a represent extremes for each stage, or more precisely, the initial tension and the arterial tension at the beginning of the next stage. Stages are treated sequentially, with finishing tensions at one step representing initial tensions for the next step, and so on. To maximize the rate of uptake or elimination of dissolved gases the *gradient*, simply the difference between p_i and p_a, is maximized by pulling the diver as close to the surface as possible. Exposures are limited by requiring that the tissue tension never exceed M, written,

$$M = M_0 + \Delta M\ d \tag{98}$$

as a function of depth, d, for ΔM the change per unit depth. A set (USN) of M_0 and ΔM are fitted below [10, 41, 70, 71] for nitrogen,

$$M_0 = 152.7\tau^{-1/4}\ fsw \tag{99}$$

$$\Delta M = 3.25\tau^{-1/4} \tag{100}$$

which can be contrasted with the original Haldane constant ratio critical tension, *M*, recognizable as the $2 - to - 1$ *law* [8, 88],

$$M = 1.58P \qquad \textbf{(101)}$$

In absolute units, the corresponding critical gradient, *G*, is given by,

$$G = \frac{M}{0.79} - P \qquad \textbf{(102)}$$

with *P* ambient pressure, and *M* critical nitrogen pressure. Similarly, the critical ratio, *R*, takes the form,

$$R = \frac{M}{P} \qquad \textbf{(103)}$$

The staging regimen is the usual,

$$p \leq M \qquad \textbf{(104)}$$

In lowest order, critical tensions for helium are the same, with helium halftimes, τ_{He}, 2.65 times faster than nitrogen halftimes, τ_{N2}, that is, by Graham's law [19, 82, 88],

$$\tau_{He} = \frac{\tau_{N_2}}{2.65} \qquad \textbf{(105)}$$

for the same compartments.

At altitude [8, 14, 27, 61, 64, 76, 82], some critical tensions have been correlated with actual testing, in which case, the depth, *d*, is often defined in terms of absolute pressure, *P*,

$$d = P - P_z \qquad \textbf{(106)}$$

with surface pressure, P_z, at altitude, *z*, given by (*fsw*),

$$P_z = 33\ exp\ (-0.0381z) = 33\alpha^{-1} \tag{107}$$

$$\alpha = exp\ (0.0381z) \tag{108}$$

and z in multiples of 1,000 feet. However, in those cases where the critical tensions have not been tested nor extended to altitude, an exponentially decreasing extrapolation scheme, called *similarity*, has been employed. Extrapolations of critical tensions, below $P = 33$ *fsw*, then fall off more rapidly then in the linear case [23, 83, 84]. The similarity extrapolation holds the ratio, $R = M/P$, constant at altitude. Denoting an equivalent sea level depth, δ, at altitude, z, one has for an excursion to depth d,

$$\frac{M(d)}{d + 33\alpha^{-1}} = \frac{M(\delta)}{\delta + 33} \tag{109}$$

so that the equality is satisfied when,

$$\delta = \alpha d \tag{110}$$

$$M(\delta) = \alpha M(d). \tag{111}$$

Considering the minimum surface tension pressure of bubbles, G^{min} (near 10 *fsw*), as a limit point, the similarity extrapolation is usually limited to 10,000 feet in elevation, and neither for decompression, nor heavy repetitive diving.

In the bulk diffusion case, depth-time exposures can be limited by a law of the form [38, 82],

$$dt_n^{1/2} = H \tag{112}$$

with t_n the nonstop time limit, and $400 \le H \le 500$ *fsw* $min^{1/2}$. One can obtain the corresponding tissue constant, λ, controlling the exposure at depth d, for nonstop time t_n, by differentiating the MTM tissue equation with respect to depth, d, and setting the result to zero. With $p_a = 0.79(d + 33)$ at sea level, there results,

$$1 - exp\ (-\lambda t_n)(1 + 2\lambda t_n) = 0. \quad \textbf{(113)}$$

Corresponding critical tensions, M, are then easily obtained using d, λ, and t_n. In the above case, the transcendental equation is satisfied when,

$$\lambda t_n = 1.25 \quad \textbf{(114)}$$

Time remaining before a stop, time at a stop, or surface interval before flying can all be obtained by inverting the tissue equation. Denoting the appropriate critical tension at some desired stage, M, and the instantaneous tension at that time, p, at stage, p_a, the time remaining, t_r, follows from,

$$t_r = \frac{1}{\lambda} \ln \left[\frac{p - p_a}{M - p_a} \right] \quad \textbf{(115)}$$

for each compartment, λ. Then, the smallest t_r controls the ascent.

The MTM is an icon in world diving circles, forming the basis for released recreational and technical dive tables, commercial protocols, and military diving. It is encoded into many Haldanian decompression meters, particularly before 1995. World Navies have performed extensive wet testing and collected data for MTM correlations and tables, using the staging paradigm of bringing the diver as close to the surface as possible. The critical tensions, M, form a springboard for neo-Haldanian models after 1990, with neo-Haldanian denoting reduced or modified M-values.

DIFFUSION MODEL (DM)

The DM dates back to the time of Hill, a contemporary of Haldane, representing an alter- native [41, 82] to bulk multitissue transfer equations with more structure imbedded. But the useful implementation of the DM dates to Hempleman [38, 60] in the 1950s. Exchange of inert gas, controlled by diffusion across regions of varying concentration, is also driven by the local gradient. Denoting the arterial blood tension, p_a, and instantaneous tissue tension, p, the gas diffusion equation takes the form in one dimensional planar geometry,

$$D\frac{\partial^2 p}{\partial x^2} = \frac{\partial p}{\partial t} \tag{116}$$

with D a single diffusion coefficient appropriate to the media. Using standard techniques of separation of variables, with ω^2 the separation constant (eigenvalue), the solution is written,

$$p - p_a = (p_i - p_a) \sum_{n=1}^{\infty} W_n \, sin \, (\omega_n x) \, exp \, (-\omega_n^2 Dt) \tag{117}$$

assuming at the left tissue boundary, $x = 0$, we have $p = p_a$, and with W_n a set of constants obtained from the initial condition. First, requiring $p = p_a$ at the right tissue boundary, $x = l$, yields,

$$\omega_n = \frac{n\pi}{l} \tag{118}$$

for all n. Then, taking $p = p_i$ at $t = 0$, multiplying both sides of the diffusion solution by $sin\ (\omega_m x)$, integrating over the tissue zone, l, and collecting terms gives,

$$W_{2n} = 0 \tag{119}$$

$$W_{2n-1} = \frac{4}{(2n-1)\pi} \tag{120}$$

Averaging the solution over the tissue domain eliminates spatial dependence, that is $sin\ (\omega_n x)$, from the solution, giving a bulk response,

$$p - p_a = (p_i - p_a) \sum_{n=1}^{\infty} \frac{8}{(2n-1)^2\pi^2} exp \, (-\omega_{2n-1}^2 Dt). \tag{121}$$

The expansion resembles a weighted sum over *effective* tissue compartments with time constants, $\omega_{2n-1}^2 D$, determined by diffusivity and boundary conditions. Unlike the perfusion case, the diffusion solution, consisting of a sum of exponentials in time, cannot be formally inverted to yield time remaining, time at a stop, nor time before flying. Such information can only be obtained by solving the equation numerically, that is, with computer or hand calculator for given M_0, ΔM, a, and b. Diffusion models fit the time constant, κ,

$$\kappa = \frac{\pi^2 D}{l^2} \tag{122}$$

to exposure data, with a typical value from the Royal Navy [38 - 40, 60] given by,

$$\kappa = 0.007928 \; min^{-1}. \tag{123}$$

The diffusion coefficient, D, however, for the above time constant, κ, above is of order $10^{-10} cm^2/sec$, five orders of magnitude slower than the aqueous tissue value, $10^{-5} cm^2/sec$, roughly. As such, it tracks morely closely the perfusion limited, multi-tissue model linked originally to Haldane [15]. The approach is aptly single tissue, with equivalent tissue halftime, τ_D,

$$\tau_D = \frac{0.693}{\kappa} = 87.5 \; min \tag{124}$$

close to the US Navy 120 minute compartment used to control saturation, decompression, and repetitive diving. Corresponding critical tensions in the bulk model, take the form,

$$M = \frac{709 \; P}{P + 404} \tag{125}$$

falling somewhere between fixed gradient and multitissue values. At the surface, $M = 53 \; fsw$, while at $200 \; fsw$, $M = 259 \; fsw$. A critical gradient, G, satisfies,

$$G = \frac{M}{0.79} - P = \frac{P\ (493 - P)}{(P + 404)}. \qquad \textbf{(126)}$$

The staging regimen is,

$$p \leq M \qquad \textbf{(127)}$$

Salient features of the bulk diffusion model can be gleaned from extension of the above slab model in the limit of thick tissue region, that is, $l \rightarrow \infty$. Replacing the summation over n with an integral as $l \rightarrow \infty$, we find

$$p - p_a = (p_i - p_a)\ e\bar{r}f\ [l/(4Dt)^{1/2}] \qquad \textbf{(128)}$$

with $e\bar{r}f$ the average value of the *error − function over l*, having the limiting form,

$$e\bar{r}f\ [l/(4Dt)^{1/2}] = 1 - (4Dt)^{1/2} l \pi^{1/2} \qquad \textbf{(129)}$$

for short times, and

$$e\bar{r}f\ [l/(4Dt)^{1/2}] = \frac{l}{(4\pi Dt)^{1/2}} \qquad \textbf{(130)}$$

for long times. The former case recovers the Hempleman [38] square root law for no decompression time *versus* depth. A variant of the model couples different *compartments* in series or parallel, assigning different properties to the compartments, and imposing flux continuity at the boundaries. Such allows for tissue-blood interface differences, and is called the n-compartment model [6, 40, 53, 54], with n the number of different tissue cells. It's been a mainstay in DCIEM [10, 53, 54] diving applications.

The DM has seen extensive testing, correlation, and use by the Royal Navy. It forms the bases of British technical and recreational dive tables, as well as other segments of the Europoean diving community at large. It was the first model employed in a mechanical analog decompression meter in the early 1970s.

SPLIT PHASE GRADIENT MODEL (SPGM)

Multitissue models address dissolved gas transport with saturation gradients driving the elimination. In the presence of free phases, free-dissolved and free-blood elimination gradients can compete with dissolved-blood gradients. One suggestion is that the gradient be split into two weighted parts, the free-blood and dissolved-blood gradients, with the weighting fraction proportional to the amount of separated gas per unit tissue volume. Use of a split phase gradient is consistent with multiphase flow partitioning, and implies that only a portion of tissue gas has separated, with the remainder dissolved. Such a split representation can replace any of the gradient terms in tissue response functions, as suggested and implemented by Wienke, Hills, and others [2, 16, 29, 42, 63, 69, 82].

Accordingly, in the split phase gradient model (SPGM), we write the rate equation, given free phase and tissue partition fractions, γ_p and γ_t, in unit tissue volume, as,

$$\frac{\partial(p - p_a)}{\partial t} = -\gamma_t \lambda (p - p_a) - \gamma_p \frac{\partial(p - P - \delta)}{\partial t} \qquad \textbf{(131)}$$

with internal pressure inside the free phase (bubbles), p_b, given by,

$$p_b = P + \delta \qquad \textbf{(132)}$$

for δ constrictive surface tension of the separated phase assembly, and P ambient pressure. The above equation cannot be inverted directly for solution, but numerical techniques can be easily applied, as the equation is a linear, first order, differential equation. Specification of free-dissolved gradient, $p - p_b$, and the surface tension, δ, requires knowledge of the dynamics of free phases in tissue and blood, or at least plausible assumptions about bubble kinetics, bubble numbers, and bubble film structures, taken up next. To schedule decompression, critical tensions, critical free phase volumes, or combinations thereof, can be employed.

The partition fractions normalize to one,

$$\gamma_p + \gamma_t = 1 \tag{133}$$

and the staging protocol is written generally [83, 88],

$$p \le \alpha M + \beta V \tag{134}$$

for constants,

$$0 \le \alpha \le 1 \tag{135}$$

$$0 \le \beta \le (1 - \alpha)\frac{M}{V_i} \tag{136}$$

with V the estimated phase volume, V_i, the nonexcited phase volume, and M the dissolved gas critical tension.

The SPGM forms a basis for neo-Haldanian decompression meters and commercial software. It can be imbedded into any MTM to yield deep stops. Deep stops are decompression stops made deeper than traditional stops required by dissolved gas models, and theoretically control and slow bubble growth. User choice of conservancy factors relate directly to α and β, for given M and estimated V and V_i. Parameters are proprietary mostly, but decompression meters employ the approach. Like the ASTM, the SPGM is found in commercially marketed software offering neo-Haldanian staging alternatives (deep and/or shallow stops) for dive planning [81].

LINEAR EXPONENTIAL MODEL (LEM)

Asymmetric uptake and elimination of dissolved gas can also be tracked by using different tissue functions for uptake and elimination. The linear-exponential model (LEM) employs exponential uptake and linear-exponential elimination tissue functions [70, 71]. The end result for tissues with the same perfusion rates is a slower outgassing rate *versus* uptake rate. That is, simply, for inert gas uptake, the rate equation is as before, with τ a characteristic tissue-blood flow time scale, not necessarily arbitrary halftimes as assigned in the MTM and other models,

$$\frac{\partial(p - p_a)}{\partial t} = -\tau^{-1}(p - p_a) \qquad \textbf{(137)}$$

with solution, $p = p_i$ at $t = 0$, for p_a arterial partial pressure,

$$(p - p_a) = (p_i - p_a)\ exp\ (-\tau^{-1}t) \qquad \textbf{(138)}$$

At some tissue tension, p_c, outgassing is *clamped*, so that the difference between the tissue and arterial tensions is constant. In that case, gas elimination on decompression is assumed to be linear until some later time in the ascent when bubbles are assumed ameliorated, and the elimination is switched back into exponential mode,

$$\frac{\partial p}{\partial t} = -\tau^{-1}(p_c - p_a) \qquad \textbf{(139)}$$

Solution, $p = p_c$ at $t = 0$, is now,

$$p = -\tau^{-1}(p_c - p_a)t + p_c \qquad \textbf{(140)}$$

The clamping is usually applied early in the decompression. In the LEM, the time constant, λ, is not necessarily linked to any MTM halftime, τ, but carries the same inverse time units. Without loss of generality, we write,

$$\lambda = \frac{1}{\tau} \qquad \textbf{(141)}$$

with τ a representative perfusion rate time parameter. One useful form is given by,

$$\lambda = \tau^{-1} = \frac{dq}{dt}\frac{\pi_b}{\pi_t} \qquad \textbf{(142)}$$

with dq/dt the blood perfusion rate in unit tissue volume, and π_t and π_b the partition fractions of tissue and blood in that same volume. For well perfused (aqueous) tissue, $\pi_b > \pi_t$, while for poorly perfused (fatty) tissue, $\pi_b < \pi_t$.

The LEM has been employed extensively in air and helium diving by the US Navy, using large data sets to correlate parameters with DCS incidence rates and risk. The LEM (and dual phase model LEPM following) have spawned probablistic approaches to diver staging [75, 76] using some of the defining model equations of the LEM and LEPM in a statistical framework fitted to data (often) with maximum likelihood. The technique was pioneered by Berghage [12] and Weathersby [80] in USN studies.

ASYMMETRIC TISSUE MODEL (ASTM)

If gas nuclei are entrained in the circulation or trapped in tissues, affecting gas transfer across boundaries and/or setting up a chain of ischemic reactions to foreign bubbles, blood perfusion rates might be lowered, an impairment with impact on gas exchange processes. This suggests a possible lengthening of tissue halftimes for elimination over those for uptake. For instance, a 10 *min* compartment for uptake becomes a 12 *min* compartment on elimination. Such lengthening procedure and the split elimination gradient render gas uptake elimination asymmetric. Instead of exponential uptake and elimination, exponential uptake and quadratic or linear elimination tissue loading functions can be used. Such modifications can be employed in any perfusion model, and tuned to the data. These approaches are used in decompression meters and dive planning software [82, 86], as well as the USN linear-exponential model (LEM) already discussed. While never measured in the laboratory, the approach will always impart conservatism to any computational algorithm [85, 88, 89].

The approach in the asymmetric tissue model (ASTM) is simple. Outgassing halftimes, τ_{out}, are slower than ingassing halftimes, τ_{in}, by inverse fractions, β, that is,

$$\tau_{out} = \beta^{-1}\tau_{in} \quad \textbf{(143)}$$

for given ingassing halftimes, τ_{in}, with,

$$0.0 < \beta \leq 1.0 \qquad \textbf{(144)}$$

Equivalently, the decay constants are linked,

$$\lambda_{out} = \beta \lambda_{in} \qquad \textbf{(145)}$$

Meters and software usually limit β roughly as,

$$0.3 \leq \beta \leq 1.0 \qquad \textbf{(146)}$$

with large decompression debts incurred in the slow compartments when $\beta \leq 0.30$, or so. The tissue equation for ingassing is the usual,

$$(p - p_a) = (p_i - p_a) \; exp \; (-\lambda_{in} t) \qquad \textbf{(147)}$$

while the outgassing equation uses outgassing constants,

$$(p - p_a) = (p_i - p_a) \; exp \; (-\lambda_{out} t) \qquad \textbf{(148)}$$

Ingassing occurs whenever,

$$p < p_a \qquad \textbf{(149)}$$

while outgassing occurs in the opposite case,

$$p > p_a \qquad \textbf{(150)}$$

Certainly, some compartments are outgassing during a dive while others are ingassing, and vice-versa. Staging maintains the tensions below critical values,

$$p \leq M \qquad \textbf{(151)}$$

Effects of slower outgassing on the ascent time are easy to quantify for single nonstop exposures. If the symmetric ascent times are denoted, t_a, then the longer (asymmetric elimination) ascent times, t_{aa}, are increased by β^{-1} at depth, d, for equal surfacing tissue tensions,

$$t_{aa} = \beta^{-1} t_a \quad \textbf{(152)}$$

While this simple relationship holds for nonstop exposures, a generalization to multilevel, mixed gas switch, repetitive, reverse profile, and decompression diving is nonexistent. The ASTM is employed in many neo-Haldanian decompression meters today. For a given set of critical tensions, M, in any MTM, the ASTM will always be more conservative because outgassing is slower. In meters [82], user conservancy knobs are controlled through β.

Dual phase diving algorithms are rather recent innovations, coming online in the past 20 years or so. They are different than dissolved gas algorithms, because they couple dissolved gases to bubbles, and lead to deeper staging as a result. Meters, tables, and software employing these algorithms do exist, and are supplanting traditional models.

THERMODYNAMIC MODEL (TM)

The thermodynamic model (TM) potentially couples both tissue diffusion and blood perfusion equations [40, 41], though the original Hills formulation with application to diving focused only on diffusion. Cylindrical symmetry is assumed in the model. From a boundary vascular zone of thickness, a, gas diffuses into the extended extravascular region, bounded by b. If we wrap the DM planar geometry into a hollow cylinder of inner radius, a, and outer radius, b, we generate Krogh geometry [41]. The hollow cylindrical model retains all the features of the planar model, and additionally includes curvature for small a and b, with $l = b - a$. Assigning the same boundary conditions at a and b, namely, the tissue tension, p, equals the arterial tension, p_a, writing the diffusion equation in radial cylindrical coordinates,

$$D\frac{\partial^2 p}{\partial r^2} + \frac{D}{r}\frac{\partial p}{\partial r} = \frac{\partial p}{\partial t} \quad \textbf{(153)}$$

and solving yields,

$$p - p_a = (p_i - p_a) \sum_{n=1}^{\infty} X_n \; U_0(\epsilon_n r) \; exp \; (-\epsilon_n^2 Dt) \quad \textbf{(154)}$$

with X_n a constant satisfying initial conditions, U_0 the cylinder functions, and α_n the eigen- values satisfying,

$$U_0(\epsilon_n a) = \frac{\partial U_0(\epsilon_n b/2)}{\partial r} = 0 \quad \textbf{(155)}$$

Averaging over the tissue region, $a \leq r \leq b$, finally gives,

$$p - p_a = (p_i - p_a)\frac{4}{(b/2)^2 - a^2}\sum_{n=1}^{\infty}\frac{1}{\epsilon_n^2}\frac{J_1^2(\epsilon_n b/2)}{J_0^2(\epsilon_n a) - J_1^2(\epsilon_n b/2)} exp \; (-\epsilon_n^2 Dt) \quad \textbf{(156)}$$

with J_1 and J_0 Bessel functions, order 1 and 0. Typical vascular parameters are bounded roughly by,

$$0 < a \leq 4 \; \mu m \quad \textbf{(157)}$$

$$10 \leq b \leq 32 \; \mu m. \quad \textbf{(158)}$$

Perfusion limiting is applied as a boundary condition through the venous tension, p_v, by enforcing a mass balance across both the vascular and cellular regions at a, according to Hennessey [40, 41],

$$\frac{\partial p_v}{\partial t} = -\kappa(p_v - p_a) - \frac{3}{a}S_p D\left[\frac{\partial p}{\partial r}\right]_{r=a} \quad \textbf{(159)}$$

with S_p the ratio of cellular to blood gas solubilities, κ the perfusion constant, and p_a the arterial tension. The coupled set relate tension, gas flow, diffusion and perfusion, and solubility in a complex feedback loop.

The thermodynamic trigger point for decompression sickness is the volume fraction, χ, of separated gas, coupled to mass balance. Denoting the separated gas partial pressure, P_{N2}, under worse case conditions of zero gas elimination upon decompression, the separated gas fraction is estimated,

$$\chi P_{N_2} = S_c(p - P_{N_2}) \quad \textbf{(160)}$$

with S_c the cellular gas solubility. The separated nitrogen partial pressure, P_{N2} is taken up by the inherent unsaturation, and given by (*fsw*),

$$P_{N_2} = P + 3.21 \quad \textbf{(161)}$$

in the original Hills [41, 42] formulation, but other estimates have been employed. Mechanical fluid injection pain, depending on the injection pressure, δ, can be related to the separated gas fraction, χ, through the tissue modulus, K,

$$K\chi = \delta \quad \textbf{(162)}$$

so that a decompression criteria requires the staging paradigm,

$$K\chi \leq \delta \quad \textbf{(163)}$$

with δ in the range, for $K = 3.7 \times 10^4 \, dyne \; cm^{-2}$,

$$0.34 \leq \delta \leq 1.13 \; fsw. \quad \textbf{(164)}$$

Identification of separated phase volume as a critical indicator is a significant development in decompression theory [39]. The TM was correlated with profiles of Australian pearl divers [49] in the Torres Straits, and was sporadically tested in the open ocean. Deep stops are central in the TM to control phase growth, but early dropout in the shallow zone often increases risk. The TM serves as a springboard for bubble models to follow.

VARYING PERMEABILITY MODEL (VPM)

Following development of the TM, Yount and co-workers [90 - 92] turned to decompressed gel studies in the laboratory as a model for bubble growth and control in diving environments. Using a critical volume of bubbles excited into growth following compression-decompression, and gel parameters for excitation radii of bubbles, the varying permeability model (VPM) was constructed. The critical radius, r_i, at fixed pressure, P_i, represents the cutoff for growth upon decompression to lesser pressure. Nuclei larger than r_i will all grow upon decompression. Additionally, following an initial compression, a smaller class of micronuclei of critical radius, r, can be excited into growth with decompression. If r_i is the critical radius at P_i, then, the smaller family, r, excited by decompression from P, obeys,

$$\frac{2\gamma}{r} - P = \frac{2\gamma}{r_i} - P_i \qquad \textbf{(165)}$$

with roughly,

$$50.0 \leq \gamma \leq 250.0 \ fsw \ \mu m \qquad \textbf{(166)}$$

for P measured in *fsw*, and r in μm. Excitation radii, r, thus follow,

$$\frac{1}{r} - \frac{1}{r_i} = \frac{P - P_i}{2\gamma} \qquad \textbf{(167)}$$

with $r_i = 0.8\ \mu m$ at $P_i = 33\ fsw$. The minimum excitation gradient, G^{min}, initially probing r, and taking into account generation of nuclei over time scales τ_r, is (fsw),

$$G^{min} = \frac{2\gamma\ (\gamma_c - \gamma)}{\gamma_c\ r(t)} = \frac{11.01}{r(t)} \qquad \textbf{(168)}$$

with,

$$r(t) = r + (r_i - r)\ [1 - exp\ (-\lambda_r t)] \quad \textbf{(169)}$$

γ, γ_c film, surfactant surface tensions, that is, γ = 17.9 *dyne/cm*, γ_c = 257 *dyne/cm*, and λ_r the inverse of the generation time for stabilized gas micronuclei (many days). Although the actual size distribution of gas nuclei in humans is unknown, experiments *in vitro* suggest that a decaying exponential is reasonable,

$$n = N\ exp\ (-\beta r) \quad \textbf{(170)}$$

with β a constant, and N a convenient normalization factor across the distribution. For small values of the argument, βr, and assuming a stabilized distribution, n_0, accommodated by the body at fixed pressure, P_0, the excess number of nuclei, Λ, excited by compression- decompression from new pressure, P, is,

$$\Lambda = n_0 - n = N\beta r_i \left[1 - \frac{r}{r_i}\right]. \quad \textbf{(171)}$$

and needs to be controlled by pressure in diving exposures, as follows. The rate at which gas grows in tissue depends upon both the excess bubble number, Λ, and the gradient, G. The critical volume hypothesis requires that the integral of the product of the two must always remain less than some limit point, $\alpha\ V$, with α a proportionality constant,

$$\int_0^\infty \Lambda G dt = \alpha V \quad \textbf{(172)}$$

for V the limiting gas volume. Assuming that gradients are constant during decompression, t_d, while decaying exponentially to zero afterwards, and taking the limiting condition of the equal sign, yields simply for a bounce dive, with λ the tissue constant,

$$\Lambda G\ (t_d + \lambda^{-1}) = \alpha V. \quad \textbf{(173)}$$

In terms of earlier parameters, one more constant, δ, closes the set, defined by,

$$\delta = \frac{\gamma_c \alpha V}{\gamma \beta r_i N} = 7180 \ fsw \ min \tag{174}$$

so that,

$$\left[1 - \frac{r}{r_i}\right] G \ (t_d + \lambda^{-1}) = \delta \frac{\gamma}{\gamma_c} = 500.8 \ fsw \ min. \tag{175}$$

The five parameters, γ, γ_c, δ, λ_r, r_i, are five of the six fundamental constants in the varying permeability model. The remaining parameter, λ_m, interpolating bounce and saturation exposures, represents the inverse time constant modulating multi-diving, while λ_r, is on the order of inverse days, that is, $\lambda_r = .0495 \ days^{-1}$. Characteristic halftimes, τ_r and τ_h, take the values $\tau_r = 14 \ days$ and $\tau_h = 12.4 \ min$. For large τ, r is close to r_i, while for small τ, r is on the order of $0.5 \ r_i$. At sea level, $r_i = 0.8 \ \mu m$. Staging on the allowed tissue supersaturation is linked to the bubble surface pressure,

$$\Pi - P \leq 2 \ \frac{\gamma}{\gamma_c r} (\gamma_c - \gamma) \tag{176}$$

with, in the permeable region,

$$r = \left[\frac{(P - P_i)}{2(\gamma_c - \gamma)} + \frac{1}{r_i}\right]^{-1} \tag{177}$$

and, in the impermeable region,

$$r^3 - 2(\gamma_c - \gamma)r^2 - \frac{P_i}{\zeta} r_c^3 = 0 \tag{178}$$

for,

$$\zeta = P - 2P_c + 2P_i + \frac{2(\gamma_c - \gamma)}{r_c} \quad \textbf{(179)}$$

according to gel measurements, with P_c near 6 *atm*. VPM parameters were also fitted to air and helium diving data in shallow zones [82]. The VPM is available in some dive planning software for technical diving [82].

REDUCED GRADIENT BUBBLE MODEL (RGBM)

The RGBM employs a phase volume constraint [82, 85, 88] across the total dive profile. The gel parameterization is replaced by flexible seed skins with appropriate EOS, permeable to gas diffusion at all pressures and temperatures. Gas diffuses across the bubble interface, and the bubble is subject to Boyle expansion-contraction. The phase volume constraint equation is rewritten in terms of a phase function, $\dot{\varphi}$, varying in time,

$$\int_0^{\tau} \frac{\partial \phi}{\partial t} dt \; \leq \; \Phi \quad \textbf{(180)}$$

with, as before,

$$\dot{\phi} = \frac{\partial \phi}{\partial t} \quad \textbf{(181)}$$

for Φ the separated phase, and τ some (long) cutoff time. More particularly, for Π the total gas tension, taking $\tau \rightarrow \infty$, holding temperature constant,

$$\dot{\phi} = \left[\frac{\partial V}{\partial t}\right]_{diffusion} + \left[\frac{\partial V}{\partial t}\right]_{Boyle} + \left[\frac{\partial V}{\partial t}\right]_{excitation} \quad \textbf{(182)}$$

for,

$$\left[\frac{\partial V}{\partial t}\right]_{diffusion} = 4\pi\beta\ exp\ (\beta\epsilon)DS\int_{\epsilon}^{\infty} nr\left[\Pi - P - \frac{2\gamma}{r}\right]dr \quad \textbf{(183)}$$

$$\left[\frac{\partial V}{\partial t}\right]_{Boyle} = 4\pi\beta\ exp\ (\beta\epsilon)\int_{\epsilon}^{\infty} nr^2\left[\frac{\partial r}{\partial P}\frac{\partial P}{\partial t}\right]dr \quad \textbf{(184)}$$

$$\left[\frac{\partial V}{\partial t}\right]_{excitation} = 4\pi\frac{\partial}{\partial t}\left[\theta(t - t_{ex})\int_{\epsilon}^{\infty} nr^2 dr\right] \quad \textbf{(185)}$$

with all quantities as denoted previously, the seed density normalized to the excited seed volume, V,

$$4\pi\int_{\epsilon}^{\infty} nr^2 dr = 4\pi\int_{\epsilon}^{\infty} exp\ (-\beta r)r^2 dr = V \quad \textbf{(186)}$$

the heaviside function, θ, defined for seed excitation at time, t_{ex},

$$\theta(t - t_{ex}) = 0, \quad t \leq t_{ex} \quad \textbf{(187)}$$

$$\theta(t - t_{ex}) = 1, \quad t > t_{ex} \quad \textbf{(188)}$$

with the time derivative of the heaviside function a delta function,

$$\frac{\partial\theta(t - t_{ex})}{\partial t} = \delta(t - t_{ex}) \quad \textbf{(189)}$$

To simplify EOS lookups, Boyle factors, ξ, are used, so that,

$$\xi PV = nRT \quad \textbf{(190)}$$

as detailed earlier. Thus the phase function, $\varphi^{'}$, depends on the number of bubbles, n, stimulated into growth by compression-decompression, the supersaturation gradient, G, seed expansion-contraction by radial diffusion, $\partial r/\partial t$, Boyle expansion-contraction with pressure changes, P, and inside temperature, T, in general. The excitation radius [16, 22, 29], α, depends on the material properties [24, 93], and is taken for nitrogen (μm),

$$\epsilon_{N_2} = 0.007655 + 0.001654 \left[\frac{T}{P}\right]^{1/3} + 0.041602 \left[\frac{T}{P}\right]^{2/3} \quad \textbf{(191)}$$

and for helium,

$$\epsilon_{He} = 0.003114 + 0.015731 \left[\frac{T}{P}\right]^{1/3} + 0.025893 \left[\frac{T}{P}\right]^{2/3} \quad \textbf{(192)}$$

for T measured in absolute ^oK, and P given in *fsw*, with ranges for virial coefficients, aqueous to lipid materials, varying by factors of 0.75 to 4.86 times the values listed above. Both expressions obtain from fits to RGBM mixed gas data across lipid and aqueous bubble films, and are different from other phase models. Values of excitation radii, ε, above range from 0.1 to 0.05 μm for sea level down to 500 *fsw*. This is compared to excitation radii in other models (VPM and TBDM) which vary in the 1 μm range. In the very large pressure limit, excitation radii (like beebees) are in the 1/1,000 μm range. Two parameters, closing the set, are nominally (STP),

$$\Phi = 840 \ \mu m^3 \quad \textbf{(193)}$$

$$\beta = 0.6221 \ \mu m^{-1} \quad \textbf{(194)}$$

fitted to collected (LANL Data Bank) deep, mixed gas, decompression data [80], and with varying bubble surface tension [26, 33],

$$2\gamma = 44.7 \left[\frac{P}{T}\right]^{1/4} + 24.3 \left[\frac{P}{T}\right]^{1/2} \ dyne/cm \quad \textbf{(195)}$$

Staging in the RGBM requires the supersaturation gradient to remain under the seed-averaged, bubble surface tension pressure,

$$\Pi - P \leq \beta \; exp \; (\beta \epsilon) \int_{\epsilon}^{\infty} exp \; (-\beta r) \frac{2\gamma}{r} dr \quad \textbf{(196)}$$

whereby the seed distribution controls the ascent.

The approach above is the basis of released mixed gas decompression tables and simplified no-group, no-calc recreational air and nitrox tables up to 10,000 ft elevation. Meter implementations of the full RGBM are available and under continuing development. The RGBM is employed in the technical, research, and scientific diving communities. It is a mainstay for LANL operations, and is used now in the commercial, oil patch, sector. Decompression meters and dive planning software are readily available with the algorithm. It has been correlated with data in the LANL Data Bank, some 2879 profiles with 20 cases of DCS with the writing of this paper. A spinoff approach, detailed next, correlates RGBM profiles for more limited application across air and nitrox recreational diving, using three bubble factors to adjust critical tensions for reproduction of full RGBM profiles across recreational diving. Bubble factors deduced from the RGBM above are employed for rapid numerical application in recreational diving meters.

TISSUE BUBBLE DIFFUSION MODEL (TBDM)

Bubbles shrink or grow according to a simple radial diffusion equation [38, 39] linking total gas tension, Π, ambient pressure, P, and surface tension, γ, to bubble radius, r,

$$\frac{\partial r}{\partial t} = \frac{DS}{r} \left[\Pi - P - \frac{2\gamma}{r} \right] \quad \textbf{(197)}$$

with D the gas diffusion coefficient, and S the gas solubility. Bubbles grow when the surrounding gas tension exceeds the sum of ambient plus surface tension pressure, and vice-versa. Higher gas solubilities and diffusivities enhance the rate. Related bubble area, A, and volume, V, changes satisfy,

$$\frac{\partial A}{\partial t} = 8\pi r \frac{\partial r}{\partial t} \tag{198}$$

$$\frac{\partial V}{\partial t} = 4\pi r^2 \frac{\partial r}{\partial t} \tag{199}$$

Using Fick's law, a corresponding molar current, J, of gas into, or out of, the bubble is easily computed assuming an ideal gas,

$$J = -\frac{DS}{RTh}\left[\Pi - P - \frac{2\gamma}{r}\right] \tag{200}$$

for R the ideal gas constant, T the temperature, and h an effective diffusion barrier thickness. And the molal flow rate is just the molal current times the interface area, that is,

$$\frac{\partial n}{\partial t} = JA \tag{201}$$

for n the number of moles of gas. The change in pressure and volume of the bubble, due to gas diffusion, follows simply from the ideal gas law,

$$\frac{\partial(PV + 2\gamma r^{-1}V)}{\partial t} = R\frac{\partial(nT)}{\partial t} \tag{202}$$

for V the bubble volume.

Obviously, the above constitute a coupled set of differential equations, solvable for a wide range of boundary and thermodynamic conditions connecting the state variables, namely, P, V, Π, r, n, and T . In the Gernhardt [35] treatment, tissue deformation, ξ, is given by,

$$\xi = \frac{4}{3}\pi r^3 Y \tag{203}$$

with Y the tissue bulk modulus, and r the bubble radius. Staging can be applied by limiting tissue deformation, such that symbolically,

$$\xi \leq \frac{4}{3}\pi r^3 Y_{min} \tag{204}$$

Bubble dose, based on the hypothetical volume of an expanding test bubble, can also be linked to staging for any exposure. Maximum likelihood regression is used to correlate bubble dose parameters with DCS risk. Staging can be then correlated with dose and risk. This approach is likely employed in the commercial diving sector [34, 82].

LINEAR EXPONENTIAL PHASE MODEL (LEPM)

Thalmann and others [70, 71, 75] have extended USN dissolved gas models to emulate bubble behavior, using both (either) modified M-values or separated bubble volumes, as critical staging parameters. The model enjoys correlation with extensive US Navy data sets accumulated over the years, and is based on statistics gathered for dissolved gas staging but not necessarily deep stops, possibly a data-model mismatch as depth and exposure time increase.

In the LEPM, the assumption that gas content remains in solution at all times during decompression is relaxed to accommodate gas bubble formation and growth. Under this relaxation, mass balance equates the rate of change of tissue compartment dissolved gas tension to the rate of gas uptake from the blood minus the gas flux into the bubbles. In essence, this is congruent with the split phase gradient model (SPGM) discussed previously. The particular form of the transfer rate equation is,

$$\frac{\partial p}{\partial t} = \frac{p_a - p}{\tau} - \frac{1}{\alpha_t V}\sum_{n=1}^{N}\frac{\partial (P_n V_n)}{\partial t} \tag{205}$$

with N the number of bubbles, P_n the inert gas pressure in the nth bubble, V_n the volume of the nth bubble, V the compartment total volume, α_t the tissue partition fraction, and τ the time constant. In the absence of bubbles, the last term vanishes, and the LEM equations are recovered.

Unlike the kinetics of exponential uptake and elimination, regardless of compartment saturation, the linear exchange kinetics of the LEM and LEPM *via* arterial-tissue tension clamping occur once bubbles are assumed to form. While this exact time is not known, the US Navy approach assumes bubble formation when a set of reduced critical tensions, denoted *MR* to distinguish them from classical M-values of multitissue models, are exceeded. The bubble growth term, $\partial V/\partial t$, has no seed dynamics imbedded in its definition (unlike RGBM and VPM), but proceeds according to an expression suggested by Van Liew [73, 74]. For r the bubble radius, and all other quantities as defined before, the growth rate is,

$$\frac{\partial V}{\partial t} = -4\pi r^2 \alpha_t D_t \left[1 - \frac{p}{p_b}\right]\left[\delta + \frac{1}{r}\right] \quad \textbf{(206)}$$

with p tissue tension, p_b reduced bubble internal pressure (active gas partial pressures subtracted from ambient pressure),

$$p_b = P - p_{H_2O} - p_{CO_2} - p_{O_2} + \frac{2\gamma}{r} \quad \textbf{(207)}$$

for D_t compartmental bulk diffusion coefficient, and δ a penetration length,

$$\delta^2 = \frac{dQ}{dt}\,\frac{\alpha_b}{\alpha_t D_t V_t} \quad \textbf{(208)}$$

within tissue volume, V_t. Boyle effects occur instantaneously with changes in ambient pressure, P, as in in other bubble models. Staging can be effected through the reduced critical tensions,

$$p \leq MR \quad \textbf{(209)}$$

as a portion of the dissolved gas is assumed to phase transition. Staging can also be effected by limiting the bubble phase volume, V_b, in time,

$$\int_0^\infty \frac{\partial V_b}{\partial t} dt = 4\pi \int_0^\infty r^2 \frac{\partial r}{\partial t} dt \leq V_{lim} \tag{210}$$

with V_{lim} the assumed limiting bubble volume, possibly inferred or deduced from data [81, 82]. Like the dissolved gas LEM, some correlations with USN data are linked to the LEPM, and the model is under continuing analyses. Statistical extensions of the model have been detailed and applied to exposure data at the Naval Experimental Diving Unit (NEDU) [70, 71, 75, 76].

FICK FOURIER EQUATION

The Fick Fourier equation can be applied to combined perfusion-diffusion applications in modeling inert gas transfer in tissue and blood, addressing both spatial and temporal evolution. The standard multitissue perfusion equations with limiters to underwater exposure times are the bases for many older dive tables, decompression meters, and software. The multitissue equations do not address spatial geometries for inert gas transfer, focusing instead on just bulk tissue response. How these two approaches map onto each other in real diving applications is interesting and important. Equivalences can be constructed and model parameters related broadly.

For decompression applications, the useful form of the Fick Fourier equation in terms of instantaneous tissue tension, p, is,

$$\frac{\partial p}{\partial t} = \nabla \cdot D \nabla p - \kappa(p - p_a) + Z \tag{211}$$

for external sources, Z, tissue diffusion constant, D, blood perfusion rate, κ, and ambient partial pressure, p_a. This comes from the usual representation in terms of concentration, c, replacing c with p from Henry's Law,

$$c = Sp \tag{212}$$

with S the solubility. Sources, Z, are zero, diffusivity, D, is locally constant, and the blood rate, K, accounts for bulk transfer. Then, we can simply write,

$$\frac{\partial p}{\partial t} = D\nabla^2 p - \kappa(p - p_a) \tag{213}$$

The first term on the RHS tracks the regional spatial evolution of inert gas while the second term on the RHS accounts for the bulk bathing of the region. The Fick Fourier equation applies to inert gas transfer in general with nitrogen and helium the usual cases. As ambient pressure, p_a, is constant, it is convenient to transform to the instantaneous difference, q,

$$q = p - p_a \tag{214}$$

so that the Fick Fourier equation reduces,

$$\frac{\partial q}{\partial t} = D\nabla^2 q - \kappa q \tag{215}$$

Temporal behavior of solutions exhibit the usual exponential decay mode in time, while the spatial solutions depend on the geometry.

Working in plane geometry is convenient and a simple representation for our purposes. The equation then takes the form,

$$\frac{\partial q}{\partial t} = D\frac{\partial^2 q}{\partial x^2} - \kappa q \tag{216}$$

Applying initial conditions in time and boundary conditions at the vascular interfaces completely defines the solution space. In planar one dimensional geometry, the Fick Fourier equation can easily be solved by standard separation of variables techniques. The solution, q, is stationary in space and time, and written in generality,

$$q = X(x)T(t) \tag{217}$$

with X and T only functions of position, x, and time, t. Substituting back into the Fick Fourier equation for planar geometry, we obtain,

$$X\frac{\partial T}{\partial t} = DT\frac{\partial^2 X}{\partial x^2} - \kappa XT \tag{218}$$

so that dividing by XT, rearranging and separating spatial and temporal terms, while equating separated operators to the same eigenvalues, α^2, the result is,

$$\frac{1}{X}\frac{\partial^2 X}{\partial x^2} = \frac{1}{DT}\frac{\partial T}{\partial t} + \frac{1}{D}\kappa = -\alpha^2 \tag{219}$$

which separately requires,

$$\frac{\partial T}{\partial t} + (\kappa + D\alpha^2)T = 0 \tag{220}$$

$$\frac{\partial^2 X}{\partial x^2} + \alpha^2 X = 0 \tag{221}$$

The time part has simple solution,

$$T = C\ exp\ [-(\kappa + D\alpha^2)t] \tag{222}$$

for C a constant. The spatial part has solution,

$$X = A\ sin\ (\alpha x) + B\ cos\ (\alpha x) \tag{223}$$

with A and B constants. The constants, A, B, C, and eigenvalues, α, are obtained from initial and boundary conditions in standard fashion. The solutions, q, take the general form,

$$q = [A\ sin\ (\alpha x) + B\ cos\ (\alpha x)]\ exp\ [-(\kappa + D\alpha^2)t] \tag{224}$$

subsuming *C* into *A* and *B* across the full vascular-tissue region for all *x*. Denoting the vascular region, *l*, with,

$$0 \leq x \leq l \tag{225}$$

and requiring $p = p_a$, or $q = 0$, at the vascular boundaries fixes eigenvalues, *α*, and integration constant, *B*, in the following way. At the inner boundary, $x = 0$, we take,

$$B = 0 \tag{226}$$

so that at the outer boundary, $x = l$, we must have,

$$\alpha_n l = n\pi \quad n = 1, 2, 3, ..., N \tag{227}$$

yielding, $p = p_a$, and $q = 0$, at inner and outer vascular boundaries. The constant, A_n, can be projected out of the expansion by multiplying both sides of the equation by $sin\ (\alpha_m x)$ and integrating over the interval, (0, *l*), at $t = 0$,

$$\int_0^l q_i\ sin\ (\alpha_m x) dx = \int_0^l \sum_{n=1}^{N} A_n\ sin\ (\alpha_n x)\ sin\ (\alpha_m x) dx \tag{228}$$

As the set, $sin\ (\alpha_n x)$, is orthogonal on the interval, (0, *l*), only $n = m$ survives in the sum. Integrating, we get,

$$q_i \frac{2}{\alpha_n} = \frac{l A_n}{2} \tag{229}$$

so that,

$$A_n = q_i \frac{4}{n\pi} \quad \textbf{(230)}$$

Collecting all terms, the solution, q, is written,

$$q = q_i \sum_{n=1}^{N} \frac{4}{n\pi} sin\ (\alpha_n x)\ exp\left[-(\kappa + D\alpha_n^2)t\right] \quad \textbf{(231)}$$

with N a convenient cutoff usually chosen with vanishing decay times. Solutions hold for the vascular region,

$$0 \leq x \leq l \quad \textbf{(232)}$$

and the tissue region,

$$l \leq x \leq \infty \quad \textbf{(233)}$$

In the vascular region, applications often use the vascular diffusivity, $D = D_{blood}$, while in the tissue regions, the tissue diffusivity is employed, $D = D_{tissue}$. The combination in the exponential rate term is denoted, Λ_n,

$$\Lambda_n = \kappa + D\alpha_n^2 \quad \textbf{(234)}$$

and consists of the perfusion rate and eigenvalues of the spatial dimension. A corresponding Fick Fourier tissue halftime, Υ_n, is defined,

$$\Upsilon_n = \frac{0.693}{\Lambda_n} \quad \textbf{(235)}$$

in the same way as the multitissue halftimes, τ_n.

The Haldane multitissue equations were cast in terms of tissue halftimes, τ_n,

$$\frac{\partial p}{\partial t} = -\lambda_n (p - p_a) \quad \textbf{(236)}$$

for,

$$\lambda_n = \frac{0.693}{\tau_n} \quad n = 1, 2, 3, ..., N \tag{237}$$

with all variables in same notation as the Fick Fourier set. Changing to the tension difference, q, again,

$$q = p - p_a \tag{238}$$

gives,

$$\frac{\partial q}{\partial t} = -\lambda_n q \tag{239}$$

with simple solutions for C an integration constant,

$$q = C\ exp\ (-\lambda_n t) \tag{240}$$

Requiring initial tension, $p = p_i$, at $t = 0$, so that, $q = q_i = p_i - p_a$, at $t = 0$, yields,

$$C = q_i \tag{241}$$

so that trivially,

$$q = q_i\ exp\ (-\lambda_n t) \tag{242}$$

In applications, halftimes, τ_n, span time domains from a few minutes to days, that is roughly,

$$2 \leq \tau_n \leq 720\ min \tag{243}$$

OXYGEN TOXICITY AND DOSE

Decompression sickness could be avoided by breathing just pure oxygen. And the usage of higher concentrations of oxygen in breathing mixtures not only facilitates metabolic function, but also aids in the washout of inert gases such as nitrogen and helium. Despite the beneficial effects of breathing oxygen at higher concentrations, oxygen proves to be toxic in excessive amounts, and over cumulative time intervals. Too little oxygen is equally detrimental to the diver. Actually, limits to oxygen partial pressures in breathing mixtures range, 0.16 *atm* to 1.6 *atm*, roughly, but symptoms of hypoxia and hyperoxia are dose dependent. Or, in other words, symptom occurrences depend on oxygen partial pressures and exposure times, just like inert gas decompression sickness. The mixed gas diver needs to pay attention not only to helium and nitrogen in staged decompression, but also cumulative oxygen exposure over the dive, and possible underexposure on oxygen depleted breathing mixtures.

The neurotoxic actions of high pressure oxygen are thought to relate directly to biochemical oxidation of enzymes, either those linked to membrane permeability or metabolic pathways. The list below is not exhaustive, but includes the following mechanisms:

- the inability of blood to remove carbon dioxide from tissue when hemoglobin is oxygen saturated;
- inhibition of enzymes and coenzymes by lipid peroxides;
- increased concentration of chemical free radicals which attack cells:
- oxidation of membranes and structural deterioration reducing electrical permeability for neuron activation:
- direct oxygen attack on smooth muscle fibres;
- oxygen induced vasoconstriction in arterioles;
- elevation of brain temperature due to lack of replacement of oxygen by carbon dioxide in hemoglobin;
- and, simple chemical kinetic redistribution of cellular carbon dioxide and oxygen with high surrounding oxygen tensions.

Fortunately for the diver, there are ways to avoid complications of hyperoxia. Careful attention to dose (depth-time) limitations for oxygen exposures is needed.

Despite the multiplicity and complexity of the above, limits for safe oxygen exposure are reasonably defined. Table **1a** below lists NOAA CNS oxygen exposure time limits, t_x, for corresponding oxygen partial pressures, p_{O2}. Below 0.5 *atm*, oxygen toxicity (CNS or pulmonary) is not really a problem. Variability in oxygen exposure limits is large, beyond variability in DCS limits for mixed gas exposures. While the 0.5 *atm* lower exposure limit is fairly well accepted, the upper exposure limits in Table **1a** are neither hard and fast nor model predictable. Oxygen M-values and tissue compartments have never been specified, quantified, nor tested. In principle, such a construction is possible from existing oxygen data.

Table 1a. Oxygen dose-time limits.

Oxygen Partial Pressure p_{O2} ***(atm)***	**Oxygen Time Limit** t_x ***(min)***	**Oxygen Tolerance (OTU)** **Y *(min)***
1.6	45	87
1.5	120	213
1.4	150	244
1.3	180	266
1.2	210	278
1.1	240	279
1.0	300	300
0.9	360	299
0.8	450	295
0.7	570	266
0.6	720	189

The data in Table **1a** is easily fitted to a dose time curve, using least squares, yielding,

$$t_x = exp\left[\frac{3.0 - p_{O_2}}{.36}\right] = 4160\ exp\ (-2.77p_{O_2})$$

or, equivalently,

$$p_{O_2} = 3.0 - .36\ ln\ (t_x)$$

in the same units, that is p_{O2} and t_x in *atm* and *min* respectively. The last column tabulates a pulmonary exposure dose, Y, for divers, called the oxygen tolerance unit (OTU), developed by Lambertsen and coworkers at the University of Pennsylvania. Formally, the oxygen tolerance, Y, is given by,

$$\Upsilon = \left[\frac{p_{O_2} - 0.5}{0.5}\right]^{0.83} t$$

and can be cumulatively applied to diving exposures according to the following prescriptions:

- maintain single dive OTUs below 1440 *min* on the liberal side, or allow for 690 *min* of that as possible full DCS recompression treatment on the conservative side, that is, 750 *min*;
- maintain repetitive total dive OTUs below 300 *min*.

The expression is applied to each and all segments of a dive, and summed accordingly for total OTUs, and then benchmarked against the 750 *min* or 300 *min* rough rule. The 750 *min* and 300 *min* OTU rules are not cast in stone in the diving community, and 10% to 25% variations are common, in both conservative and liberal directions. Formally, for multiple exposures (multilevel, deco, repetitive), the cumulative OTU, Y, is the sum of segment doses, Y_n, with segment times, t_n, and partial oxygen pressures, p_{nO2}, at each n segment,

$$\Upsilon = \sum_{n=1}^{N} \Upsilon_n = \sum_{n=1}^{N} \left[\frac{p_{nO_2} - 0.5}{0.5}\right]^{0.83} t_n$$

for N segments.

For exceptional and multiple exposures, the USN and University of Pennslyvania suggest the CNS limits summarized in (Table **1b**), where for multiple exposures, N, and segment times, t_{xn},

$$T_x = \sum_{n=1}^{N} t_{x_n}$$

Table **1b** also summarizes the depth-time relationships for oxygen.

Table 1b. Oxygen exceptional exposure time limits.

Oxygen Partial Pressure p_{O2} ***(atm)***	**Single Exposure** t_x ***(min)***	**Multiple Exposures** T_x ***(min)***
2.0	30	
1.9	45	
1.8	60	
1.7	75	
1.6	120	15
1.5	150	180
1.4	180	180
1.3	240	210
1.2	270	240
1.1	300	270
0.9	360	360
0.8	450	450
0.7	570	570
0.6	720	720

Note the severe reduction in multiple oxygen exposure time at 1.6 *atm* in Table **1b**. For this reason, technical divers generally restrict mixed gas diving exposures to $p_{O2} \leq 1.6$ *atm* throughout any sequence of dives.

A similar toxicity unit, Φ, initially introduced by Lambertsen and called the unit pulmonary toxicity dose (UPTD) is closely related to the OTU, Υ, and is given by,

$$\Phi = \left[\frac{p_{O_2} - 0.5}{0.5}\right]^{1.2} t$$

and weights oxygen partial pressures more than time in dose estimates. Both are employed in the diving community as useful oxygen depth-time limiters.

For the diver, all the foregoing translates into straightforward oxygen management protocols for both CNS and pulmonary toxicity. They are similar to inert gas management, but individual susceptibilities to oxygen seem to vary more widely, though reported statistics are more scattered. Consider CNS oxygen management first, using the CNS clock as it is popularly termed, and then pulmonary oxgen management, using the OTU as described.

• CNS Toxicity Management

The various oxygen time limits, t_x, tabulated in the Tables above, obviously bound exposures, t, at oxygen partial pressure, p_{O2}. Converting the exposure time to a fraction of the limit, Ξ_n, we can define a CNS oxygen clock, Ξ, that is over N exposure levels,

$$\Xi = \sum_{n=1}^{N} \Xi_n$$

where,

$$\Xi_n = \frac{t_n}{t_{xn}}$$

for exposure time, t_n, at level, n, with oxygen time limit, t_{xn}. Tabulating Ξ is most easily done by a computer. The prescription might be, depending on degree of conservatism,

$$0.7 \leq \Xi \leq 1.3$$

and where $\Xi = 1$ is the nominal choice. The fit equation for p_{O2} and t_x suffices to range estimates of Ξ across all depths.

For repetitive dives, a surface interval penalty, similar to the nitrogen penalty in the USN Tables, can be levied for oxygen. A 90 *min* halftime is employed today, that is, the decay constant for residual oxygen CNS management, λ_{O2}, is,

$$\lambda_{O_2} = \frac{0.693}{90} = 0.0077 \ min^{-1}$$

For surface interval, t, initial CNS clock, Ξ_i, and for 90 *min* folding time, the penalty (or residual) CNS clock, Ξ, is simply,

$$\Xi = \Xi_i \; exp\;(-0.0077t)$$

The residual value is added to the planned repetitive dive additively, just like nitrogen penalty bottom time.

• Pulmonary Toxicity Management

Pulmonary oxygen toxicity, Υ, follows a similar management scheme. As described, the total exposure, Υ, is the sum of interval exposures, Υ_n,

$$\Upsilon = \sum_{n=1}^{N} \Upsilon_n = \sum_{n=1}^{N} \left[\frac{p_{nO_2} - 0.5}{0.5}\right]^{0.83} t_n$$

and is limited,

$$300 \; min \leq \; \Upsilon \; \leq \; 750 \; min$$

depending on desired degree of conservatism, and multiplicity of repetitive dives. Variations of 15% to 25% in the exposure limits are common.

There are many ways to measure oxygen, with devices called oxygen analyzers. They are employed in chemical plants and refineries, hyperbaric chambers, intensive care units, and nurseries. The paramagnetic analyzer is very accurate, and relies on oxygen molecular response to a magnetic field in displacing inert gases from collection chambers. Thermal conductivity analyzers differentiate oxygen and nitrogen conduction properties in tracking temperatures in thermistors, with difference in temperatures proportional to the oxygen concentration. Magnetic wind analyzers combine properties of paramagnetic and thermal analyzers. Polarographic analyzers measure oxygen concentration by resistance changes across permeable oxygen membranes. Galvanic cell analyzers are micro-fuel cells, consuming oxygen on touch and generating a small current proportional to the amount of oxygen consumed. In all cases, analyzer response is linear in oxygen concentration.

Although it is tempting to avoid problems of oxygen toxicity by maintaining oxygen partial pressures, p_{O2}, far below toxic limits, this is not beneficial to inert gas elimination (free or dissolved state). Higher levels of inspired oxygen, thus correspondingly lower levels of inert gases, are advantageous in minimizing inert gas buildup and maximizing inert gas washout. Coupled to narcotic potency of helium and nitrogen, and molecular diffusion rates, balancing and optimizing breathing mixtures with decompression requirements is truly a careful technical diving exercise.

MODEL COMPLEXITIES AND CHARACTERISTICS

Having discussed bubble dynamics and models, it is worthwhile to take another look at the enormous complexity facing modelers and table designers. Models presently cannot cover all aspects of the bubble problem in divers, in fact, the whole process is stepwise and dynamically limited. The establishment and evolution of gas phases, and possible bubble trouble, involves a number of distinct, yet overlapping, steps [8, 10, 34, 41, 68, 72, 74, 78]:

- nucleation and stabilization (free phase inception);
- supersaturation (dissolved gas buildup);
- excitation and growth (free-dissolved phase interaction);
- coalescence (bubble aggregation);
- deformation and occlusion (tissue damage and ischemia).

The computational issues of bubble dynamics (formation, growth, and elimination) are mostly outside dissolved gas frameworks, but get folded into halftime specifications in a non-tractable mode. The slow tissue compartments (halftimes large, or diffusivities small) might be tracking both free and dissolved gas exchange in poorly perfused regions. Free and dissolved phases, however, do not behave the same way under decompression. Care needs be exercised in applying model equations to each component. In the presence of increasing proportions of free phases, dissolved gas equations cannot track either species accurately. Computational algorithms tracking both dissolved and free phases offer broader perspectives and expeditious alternatives, but with some changes from classical schemes. Free and dissolved gas dynamics differ. The driving force (gradient) for free phase elimination increases with depth, directly opposite to the dissolved phase elimination gradient which decreases with depth. Then, changes in operational procedures are suggested for optimality. Considerations of bubble excitation and growth invariably suggest deeper staging procedures than supersaturation methods. The gradient for free phase elimination is the difference between dissolved gas partial pressure and ambient pressure, while the gradient for free phase elimination is the difference between bubble internal pressure and ambient pressure.

Other issues concerning time sequencing of symptoms impact computational algorithms. That bubble formation is a predisposing condition for decompression sickness (DCS) is universally accepted. However, formation mechanisms and their ultimate physiological effect are two related, yet distinct, issues. On this point, most hypotheses makes little distinction between bubble formation and the onset of bends symptoms. Yet we know that *silent* bubbles [7, 8, 67] have been detected in subjects not suffering from decompression sickness. So it would thus appear that bubble formation, per se, and bends symptoms do not map onto each other in a one-to-one manner. Other factors are operative, such as amount of gas dumped from solution, size of nucleation sites receiving the gas, permissible bubble growth rates, deformation of surrounding tissue medium, and coalescence mechanisms for small bubbles into large aggregates, to name a few. These issues are the purview of bubble theories, but the complexity of mechanisms addressed does not lend itself easily to table, nor even meter, implementations. Difficulties accepted, model development and data correlation are ongoing efforts important in table fabrication, meter development, and dive planning software. The models discussed herein provide a firm foundation for extension, data updating, and broader range of application. Expect to see their continuing use and development, particularly dual phase models.

Diving computers are fairly recent developments on the diving scene, with seemingly no spikes in DCS incidence rates reported within categories of divers employing them. Yet certainly, few of these algorithms have been fully tested, particularly in the deep, decompression, and mixed gas diving zones. In time, many algorithms will be tested or analyzed with growing profile data, and protocols validated, modified, or discarded. Of course, wet and dry testing are expensive, limited in range, and not always viable operationally. In that respect, profile Data Banks with diver outcomes are enormously important to cover a full spectrum of diving not amenable nor feasible for wet and dry testing. The profile Data Banks at DAN [64] and LANL [81] are two modern ones, with DAN focused on recreational diving and LANL concerned with technical deep, mixed gas, and decompression diving. To say these Data Banks can help fill holes in the testing arena might be an understatement.

CHAPTER 2

Correlations and Validation

Abstract: Models used for safely staging divers by computers, software, or tables need correlation and validation against real diving data. The process of correlating diver profiles and DCS outcomes against predictive models and fundamental parameters is presented. Data from computer downloaded profile records is correlated against several well known models, that is, USN, ZHL16, VPM, and RGBM using maximum likelihood statistical techniques. Results are tabulated, LANL DB discussed, risk functions constructed, and implications for diving and divers detailed.

Keywords: Maximum likelihood, Model correlations, Profile data banks, Risk.

To discuss correlations and risk, we want to focus on just one model and its published results. To cover all 11 models is beyond scope of this analysis, and a modern bubble model is preferable. Accordingly, within model and data parameters, we take the LANL bubble model (RGBM), dynamical principles, and correlation with profiles in the LANL Data Bank. Table, meter, and profile risks deduced in likelihood analysis are noted along with risks parameters. The model enjoys safe, widespread, and utilitarian application in mixed gas diving, both in recreational and technical sectors, and forms the bases of software, released tables and decompression meters used by scientific, commercial, and research divers. Supercomputing power [74, 76] is employed for application and correlation of model and data. The methods and approach described are generic to all diving models.

The RGBM uses a bubble volume to limit exposures, not critical tensions. Bubble volumes are estimates of separated gas phases, and the limit point is called the phase volume. Critical tensions are limit points to dissolved gas buildup in arbitrary tissue compartments, and are often called M-values. The approach is

B.R. Wienke

computationally iterative, and though mathematically intensive, diving microprocessors today easily handle calculations in the millisecond processing time frame. The algorithm is the basis of released mixed gas technical tables [NAUI Technical Diving, Tampa, 2002] and simplified recreational air and nitrox tables up to 10,000 ft elevation. Meter implementations of the RGBM are available and under continuing development, specifically HydroSpace, Atomic Aquatics, Steam Machines, Underwater Technologies, Mares, Dacor, Suunto, ConnXcion, LiquiVision, and other players. Commercial RGBM software includes *GAP, ABYSS, Free Phase,* and HydroSpace *RGBM Simulator.* All have exhibited safe and efficient operation from diving reports.

Note so-called diving units are employed herein, that is, standard SI units for depth and pressure are not used. Pressures and depths are both measured in feet-of-seawater (*fsw*) or meters-of-seawater (*msw*). The conversion is standard,

$$10\ msw = 33.28\ fsw = 1\ atm \qquad \textbf{(244)}$$

Breathing mixtures, such as nitrox (nitrogen and oxygen), heliox (helium and oxygen), and trimix (helium, nitrogen, and oxygen), carry standardized notation. If the fraction of oxygen is greater than 21%, the mixture is termed enriched. Enriched nitrox mixtures are denoted EANx, enriched heliox mixtures are denoted EAHx, and enriched trimix mixtures are denoted EATx, for *x* the oxygen percentage. For other mixtures of nitrox and heliox the convention is to name them with inert gas percentage first, and then oxygen percentage, such as, 85/15 nitrox or 85/15 heliox. For trimix, notation is shortened to list the oxygen percentage first, and then only the helium percentage, such as, 15/45 trimix, meaning 15% oxygen, 45% helium, and 40% nitrogen. Air is interchangeably denoted EAN21 or 79/21 nitrox.

LANL PROFILE DATA BANK

Divers using modern models are reporting their profiles to a Data Bank, located at LANL (also NAUI Technical Diving Operations). The profile information requested is simple:

- bottom mix/pp_{O2}, depth, and time (square wave equivalent);
- ascent and descent rates;

- stage and decompression mix/pp_{O2}, depths, and times;
- surface intervals;
- time to fly;
- diver age, weight, and sex;
- outcome (health problems), rated 1 - 5 in order of poor (DCS) to well.

This information aids validation and extension of model application space. Some 2,879 profiles now reside in the LANL Data Bank. The are 20 cases of DCS in the data file. The underlying DCS incidence rate is, $p = 20/2879 = 0.0069$, below 1%. Stored profiles range from 150 *fsw* down to 840 *fsw*, with the majority above 350 *fsw*. All data enters through the author (BRW), that is, divers, profiles, and outcomes are filtered. A summary breakdown of DCS hit (bends) data consists of the following:

- OC deep nitrox reverse profiles – 5 hits (3 DCS I, 2 DCS II)
- OC deep nitrox – 3 hits (2 DCS I, 1 DCS II)
- OC deep trimix reverse profiles – 2 hits (1 DCS II, 1 DCS III)
- OC deep trimix – 2 hits (1 DCS I, 1 DCS III)
- OC deep heliox – 2 hits (2 DCS II)
- RB deep nitrox – 2 hits (1 DCS I, 1 DCS II)
- RB deep trimix – 2 hits (1 DCS I, 1 DCS III)
- RB deep heliox – 2 hits (1 DCS I, 1 DCS II)

DCS I means limb bends, DCS II implies central nervous system (CNS) bends, and DCS III denotes inner ear bends (occurring mainly on helium mixtures). Both DCS II and DCS III are fairly serious afflictions, while DCS I is less traumatic. Deep nitrox means a range beyond 150 *fsw*, deep trimix means a range beyond 200 *fsw*, and deep heliox means a range beyond 250 *fsw* as a rough categorization. The abbreviation OC denotes open circuit, while RB denotes rebreather. Reverse profiles are any sequence of dives in which the present dive is deeper than the previous dive. Nitrox means an oxygen enriched nitrogen mixture (including air), trimix denotes a breathing mixture of nitrogen, helium, oxygen, and heliox is a breathing mixture of helium and oxygen. None of the trimix nor heliox cases involved oxygen enriched mixtures on OC, and RB hits did not involve elevated oxygen partial pressures above 1.4 *atm*. Nitrogen-to-helium (*heavy −to −light*) gas switches occured in 4 cases, violating contemporary ICD (isobaric counterdiffusion) protocols [35, 40, 44 - 46]. Isobaric counterdiffusion refers to two inert gases (usually nitrogen and helium) moving in opposite directions in tissues and blood. When summed, total gas tensions (partial pressures) can lead to increased supersaturation and bubble formation probability. None of the set exhibited full body nor CNS (central nervous system) oxygen toxicity (popularly called *oxtox*). The 20 cases come after the fact, that is diver distress with hyperbaric chamber treatment following distress. Following are many of the profiles in the LANL Data Bank, as well as broader field testing reported to us. Profiles come from seasoned divers using wrist slate decompression tables with computer backups. Some profiles come to us directly as computer downloads, which we transcribe to the requisite format.

Profiles come from the technical diving community at large, essentially mixed gas, extended range, decompression, and extreme diving. Profiles from the recreational community are not included, unless they involve extreme exposures on air or nitrox (many repetitive dives, deeper than 150 *fsw*, altitude exposures, *etc*). This low rate makes statistical analysis difficult, and we use a global approach to defining risk after we fit the model to the data using maximum likelihood. The maximum likelihood fit links directly to the binomial probability structure of DCS incidence in divers and aviators. Consider it briefly, and the likelihood maximization technique [8, 43, 53].

PROBABILISTICS

Decompression sickness is a hit, or no hit, situation. Statistics are binary, as in coin tossing. Probabilities of occurrence are determined from the binomial distribution, which measures the numbers of possibilities of occurrence and non-occurrence in any number of events, given the incidence rate. Specifically, the probability, P, in a random sample of size, N, for n occurrences of decompression sickness and m non-occurrences, takes the form,

$$P(n) = \frac{N!}{n!\ m!}\ p^n q^m \qquad \textbf{(245)}$$

with,

$$n + m = N \quad \textbf{(246)}$$

p the underlying incidence rate (average number of cases of decompression sickness), and *q*,

$$q = 1 - p \quad \textbf{(247)}$$

the underlying non-incidence. For large sample sizes, $N = n + m$,

$$ln\ P(n) \approx N\ ln\ N - n\ ln\ n - m\ ln\ m + n\ ln\ p + m\ ln\ q \quad \textbf{(248)}$$

The likelihood of binomial outcome, Φ, of *N* trials is the product of individual measures of the form,

$$\Phi(n) = p^n q^m = p^n (1 - p)^m \quad \textbf{(249)}$$

given *n* cases of decompression sickness and *m* cases without decompression sickness, and,

$$n + m = N \quad \textbf{(250)}$$

The natural logarithm of the likelihood (LL), Ψ, is easier to use in applications,

$$\Psi = \ln \Phi = n \ln\ p + m \ln\ (1 - p) \quad \textbf{(251)}$$

and is maximized when,

$$\frac{\partial \Psi}{\partial p} = 0 \quad \textbf{(252)}$$

Multivalued probability functions, $p(x)$, generalize in the maximization process,

$$\frac{\partial \Psi}{\partial p} = \sum_{k=1}^{K} \frac{\partial \Psi}{\partial x_k} \frac{\partial x_k}{\partial p} = 0 \quad \textbf{(253)}$$

satisfied when,

$$\frac{\partial \Psi}{\partial x_k} = 0 \ \ for \ k = 1, \ K \quad \textbf{(254)}$$

In application, such constraints are most easily solved on computers, with analytical or numerical methods. For RGBM analysis, the likelihood, Ψ, is typically a function of 2 - 3 parameters over the whole set of profiles. This requires extensive computing power coupled to sophisticated numerical techniques and software.

Hundreds of air dives were analyzed using this procedure, permitting construction of decompression schedules with 95% and 99% non incidence (5% and 1% bends incidence). Tables were published by US Navy investigators [5, 73], and Table **2** tabulates the corresponding non- stop time limits (p = 0.05, 0.01), and also includes the standard US Navy (Workman) limits [21, 23, 86] for comparison. Later re-evaluations of the standard set of nonstop time limits estimate a probability rate of 1.25% for the limits. In practice, incidence rates are below 0.001%, and most divers do not dive to the limits.

Table 2. Nonstop time limits for 1% and 5% DCS probability.

Depth ***d (fsw)***	**Nonstop Limit** t_n ***(min)*** $p = .05$	**Nonstop Limit** t_n ***(min)*** $p = .01$	**Nonstop Limit** t_n ***(min)*** **US Navy**
30	240	170	
40	170	100	200
50	120	70	100
60	80	40	60
70	80	25	50
80	60	15	40
90	50	10	30

(Table 2) contd.....

Depth d (*fsw*)	Nonstop Limit t_n (*min*) $p = .05$	Nonstop Limit t_n (*min*) $p = .01$	Nonstop Limit t_n (*min*) US Navy
100	50	8	25
110	40	5	20
120	40	5	15
130	30	5	10

LANL DATA CORRELATIONS AND RISK ESTIMATORS

To perform risk analysis with the LANL Data Bank, an estimator need be selected. For diving, dissolved gas and phase estimators are useful. Two, detailed earlier, are extended here. First is the dissolved gas supersaturation ratio, historically coupled to Haldane models, ρ, written in modified ratio form,

$$\rho(\kappa, \lambda, t) = \kappa \left[\frac{\Pi(t) - P(t)}{P(t)} \right] - \kappa \; exp \; (-\lambda t) \tag{255}$$

and second, ψ, is the separated bubble volume, employed by dual phase models,

$$\psi(\gamma, \mu, t) = \gamma \left[\frac{\phi(t)}{\phi_i(t)} \right] - \gamma \; exp \; (-\mu t) \tag{256}$$

with $\varphi(t)$ the bubble volume due to excitation, diffusion, and Boyle expansion-contraction, and φ_i the initial bubble excitation volume. The exponential terms in both risk functions merely insure data smoothing for short dives, that is, as $t \rightarrow 0$, then $r \rightarrow 0$, too. For long dives, $t \rightarrow \infty$, the exponential terms vanish. Physically, the exponential terms also link to bubble extinction, not discussed herein. Both risk functions vary in time, exposure, and staging. For simplicity, the asymptotic exposure limit is used in the likelihood integrals for both risk functions, r, across all compartments, τ,

$$1 - r(\kappa, \lambda) = exp \left[- \int_0^\infty \rho(\kappa, \lambda, t) dt \right] \tag{257}$$

$$1 - r(\gamma, \mu) = exp\left[-\int_0^\infty \psi(\gamma, \mu, t)dt\right] \quad \textbf{(258)}$$

with *hit – no hit*, likelihood function, Ω, of form,

$$\Omega = \prod_{k=1}^{K} \Omega_k \quad \textbf{(259)}$$

$$\Omega_k = r_k^{\delta_k}(1 - r_k)^{1-\delta_k} \quad \textbf{(260)}$$

and logarithmic reduction, Ψ,

$$\Psi = ln\ \Omega \quad \textbf{(261)}$$

where, $\delta_k = 0$ if DCS does not occur in profile, *k*, or, $\delta_k = 1$ if DCS does occur in profile, *k*. To estimate κ, λ, γ, and μ in maximum likelihood, a modified Levenberg-Marquardt [6, 43] algorithm is employed (*SNLSE*, Common Los Alamos Applied Mathematical Software Library) [84], a nonlinear least squares data fit (NLLS) to an arbitrary logarithmic function (minimization of variance over *K* data points with *L*2 error norm). The same technique was applied to estimating separated phase volume and inert gas densities.

We assign numerical tasks to processors on the LANL Blue Mountain Machine, a massively parallel processor (MPP) with 2,000 nodes according to:

- each tissue compartment, τ, then, within each compartment;
- only nitrox data points;
- only trimix data points;
- only heliox data points;
- both nitrox and trimix data points;
- both nitrox and helium data points;
- both heliox and trimix data points;
- all heliox, nitrox, and trimix data points.

estimating κ, λ, γ, and μ across all domains. The last case, all data, is the full set employed in risk analysis, but there wasn't much difference in the estimators, seen in mean error estimates across the partitioned data structures. For 11 tissue compartments, and 7 data sets, 77 risk estimates emerge. Only maximum tissue risks are finally averaged and variance computed. In diver staging, certain tissue compartments control the exposure, This is true within dis- solved gas algorithms, as well as bubble algorithms. Finally, we find across the partitioned data structures, 2-8 above:

$$\kappa = 0.91 \pm 0.14\ min^{-1} \quad \textbf{(262)}$$

$$\lambda = 0.28 \pm 0.11\ min^{-1}. \quad \textbf{(263)}$$

and, similarly.

$$\gamma = 0.09 \pm 0.07\ min^{-1} \quad \textbf{(264)}$$

$$\mu = 0.88 \pm 0.46\ min^{-1} \quad \textbf{(265)}$$

For notational shorthand, we abbreviate supersaturation and bubble risk functions,

$$\sigma = r(\kappa, \lambda) \quad \textbf{(266)}$$

$$\beta = r(\gamma, \mu) \quad \textbf{(267)}$$

LOGARITHMIC LIKELIHOOD AND SIGNIFICANCE

The data is relatively coarse grained, making compact statistics difficult. The incidence rate across the whole set is small, on the order of 1% and smaller. Fine graining into depths is not meaningful yet, so we breakout data into gas categories (nitrox, heliox, trimix), as tabulated earlier. Table **3a** indicates the breakdown.

The DCS hit rate with nitrox is higher, but not statistically meaningful across this sparse set. The last entry is all mixes, as noted previously. In the above set, there are 35 *marginals* not included in the likelihood calculations.

Table 3a. Profile data.

Mix	Total Profiles	DCS Hits	Incidence
OC nitrox	344	8	0.0232
RB nitrox	550	2	0.0017
all nitrox	894	10	0.0112
OC trimix	656	4	0.0061
RB trimix	754	2	0.0027
all trimix	1410	6	0.0042
OC heliox	116	2	0.0172
RB heliox	459	2	0.0044
all heliox	575	4	0.0070
all	2879	20	0.0069

The logarithmic likelihood (LL), Ψ, is a rough metric for fits to bubble and supersaturation risk estimators. The canonical value, Ψ_6, is the LL for the 6 OC/RB gas control data set. No fit value, Ψ, will better the canonical value, Ψ_6, that is,

$$\Psi_6 = -112.9 \quad \textbf{(268)}$$

$$\Psi \leq \Psi_6 \quad \textbf{(269)}$$

meaning all fits will be more negative (smaller LL). Results are tabulated as follow in Table **3b**.

Table 3b. Logarithmic likelihood and logarithmic likelihood ratio.

Estimator	LL	Parameters	LLR	α
6 step set	$\Psi_6 = 112.9$	$p = 0.0232, 0.0061, 0.0172,$ $0.0036, 0.0027, 0.0044$		
3 step set	$\Psi_3 = -118.4$	$p = 0.0112, 0.0042, 0.0079$	$\Gamma_3 = 11.0$	0.013
1 step set	$\Psi_1 = -119.2$	$p = 0.0069$	$\Gamma_1 = 12.6$	0.033
σ	$\Psi_{sat} = -210.6$	$\kappa = 0.91 \pm 0.14\ \text{min}^{-1}$ $\lambda = 0.28 \pm 0.11\ \text{min}^{-1}$	$\Gamma_{sat} = 92.2$	0.001

(Table 3b) contd.....

Estimator	LL	Parameters	LLR	α
β	$\Psi_{bub} = -113.3$	$\gamma = 0.09 \pm 0.07$ min^{-1} $\mu = 0.88 \pm 0.46$ min^{-1}	$\Gamma_{bub} = 0.8$	0.933

The logarithmic likelihood ratio (LLR), denoted Γ, tests two models, and is χ^2 distributed,

$$\Gamma = 2(\Psi_6 - \Psi) \tag{270}$$

for Ψ the bubble and supersaturation estimators in Table **3b**. The percentage point, α, is the area under the χ^2 curve, from $\chi^2_{a,v} = \Gamma$ to ∞,

$$\int_{\chi^2_{\alpha,\nu}}^{\infty} \chi^2(x,\nu)dx = \alpha \tag{271}$$

for v the degrees of freedom (6 - the number of bubble, supersaturation, 3 step, or 1 step degrees of freedom). The *hit − no hit* criteria for the bubble estimator is the phase volume, Φ, while standard M-values are the criteria for the supersaturation estimator. Deep stops clobber traditional M-values.

Clearly, the supersaturation risk function does not correlate well, compared to the bubble risk function. It does not work here in the deep decompression arena, but others [73, 74] have shown it correlates in the nonstop, or light decompression, limits. In those limits, bubble models and supersaturation models tend to converge, simply because phase growth is minimal.

POPULAR MODEL CORRELATIONS WITH DEEP STOP DATA

A diving *protocol* is a combination of model, data and ascent staging procedure that can be safely used across commercial, sport, technical, research, and scientific underwater operations [15, 38, 68, 84]. Accordingly, this work analyzes four popular models against actual diver deep stop profile data and DCS outcomes. Only computer downloaded profiles with DCS outcomes are considered in this analysis. The models are the USN [85], ZHL16 [15], VPM [89], and RGBM [75]. The USN and ZHL16 models are dissolved gas models [1] that ultimately require decompression stops in the shallow zone to eliminate dissolved nitrogen and helium. The VPM and RGBM are coupled bubble-dissolved gas models [77] that require deeper decompression stops to control bubble growth and dissolved gas elimination. The efficiency of shallow stops *versus* deep stops is one of current interest [4, 48, 74], and this study quantifies the utility of coupled deep stop model and data as a useful and safe diver staging tool. Many protocols are based on shallow stop data which focus on just dissolved gas buildup and elimination. Both are used today. The shallow stop models, USN and ZHL16, have been tested in man trials over the years, while the deep stop models, VPM and RGBM, have not yet.

Collecting real world diving data is a global alternative to differential wet and dry testing, a very precise but limited statistical procedure. The approach here for technical, mixed gas, and deep decompression diving parallels the Project Dive Exploration (PDE) and Diving Safety Laboratory (DSL) efforts at DAN [16] for recreational air and nitrox diving, but does not overlap significantly. As will be seen, the deep stop models (VPM, RGBM) correlate well with the LANL DB, while the shallow stop models (USN, ZHL16) do not. Turns out that both shallow and deep stops can be made at the same relative risk level, but deep stops usually admit shorter overall decompression times [76, 89], an important aspect of operational diving when mission objectives are folded over diving requirements, especially diver safety.

To correlate hundreds of *g bytes* of downloaded dive computer data (commercial and specialized meters) supercomputing power here at LANL is advantageous in performing maximum likelihood analyses of data and model. Using powerful software, the transitions from microprocessors to parallel processors are seamless. The software package, *CLAMSL* (Common Los Alamos Mathematical Statistical Library [74], distributed across all processors both parses the data and performs calculations using the most appropriate statistical methods. Little user intervention is requisite and data problems are tagged as user flags if any occur. Run times for the full up analysis (data loading, correlation, risk function fitting) on the Blue Mountain Massively Parallel Processor (MPP) are of order 40 *sec*. Peak processing speeds of 3.1 *tf lops* suggest (minimally) some 12.4 *pflop* over the full analysis. World fastest PCs operate at 374 *g flops* and, if able to even compile massive mathematical libraries and related software, would exhibit corresponding run times of 5+ *hrs*. Nominally, a 2.5 *g hertz* chip turns 10 *gflops*.

All models use tissue perfusion equations for dissolved gas buildup and elimination. Denoting instantaneous tension, p, for initial tissue tension, p_i, at ambient pressure, p_a, with tissue halftime, τ, we have,

$$p - p_a = (p_i - p_a)\ exp\ (-\lambda t) \tag{272}$$

at exposure time, t, and,

$$\lambda = \frac{0.693}{\tau} \tag{273}$$

Sets of tissue halftimes, τ, vary across models but differences are not important *for this analysis*, with an approximate range,

$$1 \leq \tau \leq 720\ min \tag{274}$$

Tissues with small τ are termed fast, while tissues with large τ are termed slow. Helium compartments are roughly 3 times faster than nitrogen compartments. In mixtures of inert gases (nitrogen and helium usually), the total tissue tension, Π, is the sum over mixture components,

$$\Pi = \sum_{j=1}^{N} p_j \tag{275}$$

with p_j the tension of the jth gas component and N the number of gas components in the mixture, usually just nitrogen and helium. After this, models diverge in their diver staging regimens, with dissolved gas models (USN and ZHL16) limiting dissolved gas buildup, Π, and bubble models (VPM and RGBM) coupling dissolved gas buildup to bubble growth and limiting bubble volumes, Φ. In all models, a permissible supersaturation, G, can be defined at each point of the dive, and is the parameter that will be correlated with data and discussed next for each of the four models.

1. USN Model (Workman 1965)

In the Workman USN approach, the permissible gas tension, Π, is limited by,

$$\Pi \leq M \tag{276}$$

with *M* critical tensions listed in Table **4a** for depth, *d*,

$$M = M_0 + \Delta M d \tag{277}$$

where depth, *d*, is the difference between total ambient pressure, *P*, and surface pressure, *P*0,

$$d = P - P_0 \tag{278}$$

Table 4a. USN M-Values.

Nitrogen			Helium		
τ_{N2} *(min)*	M_0 *(fsw)*	ΔM	$^{\tau}He$ *(min)*	M_0 *(fsw)*	ΔM
5	104	1.8	5	86	1.5
10	88	1.6	10	74	1.4
20	72	1.5	20	66	1.3
40	56	1.4	40	60	1.2
80	54	1.3	80	56	1.2
120	52	1.2	120	54	1.2
160	51	1.1	160	54	1.1
200	51	1.1	200	53	1.0
240	50	1.1	240	53	1.0

Corresponding permissible gradients, *G*, then satisfy,

$$G = \Pi - P \leq M - P = (M_0 - \Delta M P_0) + (\Delta M - 1)P \tag{279}$$

with P_0 ambient pressure at the surface as noted,

$$P_0 = 33\ exp\ (-0.0381h) \qquad \textbf{(280)}$$

for elevation, *h*, in multiples of 1000 *ft*. The correlation parameter, χ, is,

$$\chi = G - (M - P) \qquad \textbf{(281)}$$

with surfacing hit criteria,

$$\Pi > M_0 \qquad \textbf{(282)}$$

2. ZHL16 Model (Buhlmann 1990)

The Buhlmann ZHL16 approach is similar to the Workman USN approach, that is, the permissible gas tension, Π, is limited by,

$$\Pi \leq Z \qquad \textbf{(283)}$$

with critical tensions, *Z*, given by,

$$Z = a + \frac{P}{b} = a + \frac{P_0 + d}{b} \qquad \textbf{(284)}$$

so that,

$$G = \Pi - P \leq a + \left[\frac{1}{b} - 1\right] (P_0 + d) \qquad \textbf{(285)}$$

for constants, *a* and *b*, defining *Z* at sea level, in Table **4b**.

Table 4b. Buhlmann Z-Values nitrogen helium.

Nitrogen			Helium		
τ_{N2} (*min*)	$a + 33/b$ (*fsw*)	$1/b$	τ_{He} (*min*)	$a + 33/b$ (*fsw*)	$1/b$
4.0	106.2	1.91	1.5	134.5	2.36
8.0	83.2	1.54	3.0	102.4	1.74
12.5	73.8	1.39	4.7	89.4	1.53
18.5	66.8	1.28	7.0	79.8	1.38
27.0	62.3	1.23	10.2	73.6	1.32
38.3	58.4	1.19	14.5	68.2	1.25
54.3	55.2	1.15	20.6	63.7	1.21
77.1	52.3	1.12	29.0	59.7	1.17
109.2	49.8	1.09	41.1	57.1	1.14
146.0	48.2	1.08	55.2	55.1	1.12
187.0	46.8	1.07	70.7	54.0	1.11
239.0	45.6	1.06	90.3	53.3	1.10
305.0	44.5	1.05	115.3	53.1	1.09
390.0	43.5	1.04	147.4	52.8	1.09
498.0	42.6	1.04	188.2	52.6	1.08
635.0	41.8	1.03	240.0	52.3	1.07

The correlation parameter, χ, is,

$$\chi = G - (Z - P) \qquad \textbf{(286)}$$

with hit criteria at the surface,

$$\Pi > a + \frac{P_0}{b} \qquad \textbf{(287)}$$

3. Varying Permeability Model (Yount 1986)

The tissue compartments in the Yount VPM for nitrogen consist of the set,

$$\tau_{N_2} = (1, 2, 5, 10, 20, 40, 80, 120, 160, 240, 320, 400, 480, 560, 720)\ min \quad \textbf{(288)}$$

with the helium compartments scaling,

$$\tau_{He} = \frac{\tau_{N_2}}{3} \quad \textbf{(289)}$$

The VPM model links to bubble experiments in gels and related strata. In gel experiments, Yount divided gas diffusion across bubble interfaces into permeable and impermeable regions. For dive model applications, the regions separate around 165 *fsw*. Bubbles of nitrogen and helium are excited into growth by pressure changes during the dive from some minimum excitation radius, ε, in the 0.5 μm range, with nitrogen bubbles slightly larger than helium bubbles and the excitation radius decreasing with increasing absolute pressure, P. The excitation radius separates growing from shrinking bubbles. The radial bubble distribution, n, in the VPM is given by,

$$n = n_0\ exp\ (-\beta r) \quad \textbf{(290)}$$

with n_0 an experimental normalization factor for gel sample size, and β on the order of $1/\varepsilon\ \mu m^{-1}$ for diving applications. The staging protocol in the VPM limits the permissible supersaturation, G to prevent bubble growth on ascent,

$$G = \Pi - P \leq \frac{\gamma}{\gamma_c}\left[\frac{2\gamma_c}{\epsilon} - \frac{2\gamma}{\epsilon_0}\right] \quad \textbf{(291)}$$

with γ the usual bubble surface tension, and γ_c the crushing bubble effective surface tension, roughly 20 *dyne/cm* and 150 *dyne/cm* respectively [23]. The radius, ε_0, is an experimental metric, somewhere near 0.7 μm. For diving, VPM ascents are limited by G at each stage in the decompression and staging profiles are iterated to convergence across all stops. The correlation parameter, χ, in the VPM is,

$$\chi = G - \frac{\gamma}{\gamma_c}\left[\frac{2\gamma_c}{\epsilon} - \frac{2\gamma}{\epsilon_0}\right] \quad \textbf{(292)}$$

and surfacing hit condition,

$$\Pi > P_0 + \frac{\gamma}{\gamma_c}\left[\frac{2\gamma_c}{\epsilon} - \frac{2\gamma}{\epsilon_0}\right] \quad \textbf{(293)}$$

4. Reduced Gradient Bubble Model (Wienke 2008)

Nitrogen tissue compartments in the Wienke RGBM range,

$$\tau_{N_2} = (2, 5, 10, 20, 40, 80, 120, 160, 200, 240, 300)\ min \quad \textbf{(294)}$$

with helium compartments,

$$\tau_{He} = \frac{\tau_{N_2}}{2.65} \quad \textbf{(295)}$$

using the ratio of the square root of atomic weights as the scaling factor. The bubble dynamical protocol in the RGBM model amounts to staging on the seed number averaged, free-dissolved gradient across all tissue compartments, G, for P permissble ambient pressure, Π total inert gas tissue tension, n excited bubble distribution in radius (exponential), γ bubble surface tension, and r bubble radius,

$$G\int_{\epsilon}^{\infty} ndr = (\Pi - P)\int_{\epsilon}^{\infty} ndr \leq \int_{\epsilon}^{\infty}\left[\frac{2\gamma}{r}\right] ndr \quad \textbf{(296)}$$

so that,

$$G = (\Pi - P) \leq \beta\ exp\ (\beta\epsilon)\int_{\epsilon}^{\infty} exp\ (-\beta r)\left[\frac{2\gamma}{r}\right] dr \quad \textbf{(297)}$$

for ε the excitation radius at P. Time spent at each stop is iteratively calculated so that the total separated phase, Φ, is maintained at, or below, its limit point. This requires some computing power, but is attainable in diver wrist computers presently marketed, with the same said for the VPM. The USN and ZHL16 models are less complex for computer implementation. The limit point to phase separation, Φ, is near 600 μm^3, and the distribution scaling length, *β*, is close to 0.60 μm^{-1} for both nitrogen and helium. Both excitation radii, *ε*, and surface tension, *γ*, are functions of ambient pressure and temperature, and not constant. The equation-of-state (EOS) assigned to the bubble surface renders the surface tension below lipid estimates, on the order of 15 *dyne/cm*, and excitation radii are below 1 *μm* [18]. Correlation parameter, *χ*, in the RGBM is given by,

$$\chi = G - \beta \ exp \ (\beta \epsilon) \int_{\epsilon}^{\infty} exp \ (-\beta r) \left[\frac{2\gamma}{r} \right] dr \quad \textbf{(298)}$$

and surfacing hit criteria has,

$$\Pi > P_0 + \beta \ exp \ (\beta \epsilon) \int_{\epsilon}^{\infty} exp \ (-\beta r) \left[\frac{2\gamma}{r} \right] dr \quad \textbf{(299)}$$

Rigorous statistical techniques can be applied to binomial data for arbitrary values of the underlying incidence, *p*. While large values of *p* might be desirable for laboratory (controlled) wet and dry testing, they are not desirable for real divers submitting profiles to DBs in general. The underlaying incidence in the DAN and LANL DBs is less than 1%. The Weibull function is useful for fitting applications where $p \leq 0.01$ and used herein. Computer records from real divers are likely to stay below incidences rates of 1%.

To perform risk analysis with the LANL DB, an estimator need be selected. For both dissolved gas and phase models the permissible supersaturation, *G*, is useful. As detailed earlier and discussed elsewhere [84], the permissible supersaturation, *G*, is cast into normalized risk function, *ρ*, form,

$$\rho(\kappa, \omega, t) = \kappa \left[\frac{\Pi(t) - P(t)}{P(t)} \right] - \kappa \ exp \ (-\omega t) \quad \textbf{(300)}$$

with Π(*t*) and *P* (*t*) total tissue tension and ambient pressures in time, *t*. the exponential term, *exp* (*ωt*), damps out early unimportant scatter in the profiles. The asymptotic exposure limit is used in the likelihood integrals for risk function, *r*, across all compartments, *τ*,

$$1 - r(\kappa, \omega) = exp\left[-\int_0^\infty \rho(\kappa, \omega, t)dt\right] \quad \textbf{(301)}$$

with *hit* − *no hit*, likelihood function, Ω, of form,

$$\Omega = \prod_{k=1}^{K} \Omega_k \quad \textbf{(302)}$$

$$\Omega_k = r_k^{\delta_k}(1 - r_k)^{1-\delta_k} \quad \textbf{(303)}$$

and logarithmic reduction, Ψ,

$$\Psi = ln\ \Omega \quad \textbf{(304)}$$

where, $\delta_k = 0$ if DCS does not occur in profile, *k*, or, $\delta_k = 1$ if DCS does occur in profile, *k*. To estimate κ and ω in maximum likelihood, a modified Weibull-Levenberg-Marquardt [52] model is employed (*SNLSE*, Common Los Alamos Mathematical Statistical Library), a nonlinear least squares data fit (NLLS) to an arbitrary logarithmic function (minimization of variance over *K* data points with *L*2 error norm).

We assign numerical tasks to the 2000 processors on the LANL Blue Mountain MPP according to tissue compartments and the 6 (nitrox, trimix, heliox) data sets. Risk estimates emerge and risk parameters are finally averaged and variance computed. In diver staging, certain tissue compartments control the exposure, This is true within dissolved gas models, as well as bubble models. These compartments change on ascent from fast to slower compartments. In applications from nonstop to decompression diving in the LANL DB, controlling compartments, τ_c, range roughly,

$$2 \leq \tau_c \leq 300\ min \quad \textbf{(305)}$$

and long, prolonged, saturation diving is not a profile characteristic. It is these compartments at various points in the ascent that are correlated, as the others do not control the ascent at that point in space and time.

Finally, we fit across the 6-step nitrox, heliox, trimix data set::

$$\text{USN} - \kappa = 0.45 \pm 0.16\ min^{-1},\ \omega = 0.82 \pm 0.09\ min^{-1}$$
$$\text{ZHL16} - \kappa = 0.56 \pm 0.23\ min^{-1},\ \omega = 0.89 \pm 0.16\ min^{-1}$$
$$\text{VPM} - \kappa = 0.83 \pm 0.17\ min^{-1},\ \omega = 1.02 \pm 0.29\ min^{-1}$$
$$\text{RGBM} - \kappa = 0.96 \pm 0.13\ min^{-1},\ \omega = 0.91 \pm 0.18\ min^{-1}$$

The logarithmic likelihood (LL), Ψ, is a rough metric for fits to bubble and supersaturation risk estimators. The canonical value, Ψ_6, is the LL for the 6-step data set. No fit value, Ψ, will better the canonical value, Ψ_6, that is,

$$\Psi_6 = -124.86 \quad \textbf{(306)}$$

$$\Psi \le \Psi_6 \quad \textbf{(307)}$$

meaning all fits will be more negative (smaller LL). Results are tabulated as follow in Table **4c**. The 6-step set, nitrox OC and RB, heliox OC and RB, and trimix OC and RB profiles, is non-sparse and is the same set employed in previous analyses. The 3-step set is all nitrox, heliox, and trimix profiles across all depths and breathing systems. The 1-step set is just all profiles across depths, breathing mixtures, and breathing systems.

The logarithmic likelihood ratio (LLR), denoted Γ, tests two models, and is χ^2 distributed,

$$\Gamma = 2(\Psi_6 - \Psi) \quad \textbf{(308)}$$

for Ψ the bubble and supersaturation estimators in Table **4c**. The percentage point, α, is the area under the chi squared curve, from Γ to ∞, measuring the goodness of fit, ranging 0 to 1, and denoted in customary fashion,

$$\int_{\Gamma}^{\infty} \chi^2(x,\nu)dx = \alpha \quad \textbf{(309)}$$

with,

$$\chi^2(x,\nu) = \frac{1}{2^{\nu/2}\gamma\left(\nu/2\right)} x^{\nu/2-1} exp\left(-x/2\right) \quad \textbf{(310)}$$

with γ (ν/2) the complete gamma function for ν the degrees of freedom (6 minus the number of USN, ZHL16, VPM, RGBM, 3, or 1-step degrees of freedom). Here, ν will vary between 5 and 3, that is, for the USN, ZH16, VPM, and RGBM correlations, ν = 4, while for the 3-step correlation, ν = 3, and for the 1-step correlation, ν = 5. Standard software is available to estimate α for given Γ and ν and used here.

It is important to note that specific model parameters are fixed herein, and not fitted in likelihood analysis. Such would be a separate study, which using the 6-step set, would allow up to 5 model values to be optimized. Nominal model parameters were briefly mentioned and annotated earlier in the discourse. Data correlations with nominal model parameter settings are of general interest across the diving community and our focus here.

Table 4c. Logarithmic likelihood and logarithmic likelihood ratio.

Estimator	LL	Parameters	LLR	α
6 step set	$\Psi_6 = -124.86$	p = 0.0095, 0.0044, 0.0061, 0.0140,0.1818,0.5000	$\Gamma_6 = 0$	1.000
3 step set	$\Psi_3 = -133.71$	p = 0.0112, 0.0054, 0.0080	$\Gamma_3 = 16.7$	0.031
1 step set	$\Psi_1 = -134.86$	p = 0.0077	$\Gamma_1 = 20.0$	0.027
USN	$\Psi_{USN} = -133.0$	$\kappa = 0.45 \pm 0.16\ min^{-1}$ $\omega = 0.82 \pm 0.09\ min^{-1}$	$\Gamma_{USN} = 8.3$	0.081
ZHL16	$\Psi_{ZHL16} = -128.3$	$\kappa = 0.56 \pm 0.23\ min^{-1}$ $\omega = 0.89 \pm 0.16\ min^{-1}$	$\Gamma_{ZHL16} = 7.1$	0.132
VPM	$\Psi_{VPM} = -126.9$	$\kappa = 0.83 \pm 0.17\ min^{-1}$ $\omega = 1.02 \pm 0.29\ min^{-1}$	$\Gamma_{VPM} = 2.1$	0.717
RGBM	$\Psi_{RGBM} = -126.1$	$\kappa = 0.96 \pm 0.13\ min^{-1}$ $\omega = 0.91 \pm 0.18\ min^{-1}$	$\Gamma_{RGBM} = 1.3$	0.861

The USN, ZHL16, VPM, and RGBM models and protocols were statistically correlated with profiles in the LANL DB. The DB stores technical, decompression, mixed gas diving profiles with outcomes. Some 2879 computer downloaded, deep stop, profiles reside within the DB, with 20 cases of DCS. Nominal model user parameters were employed in calculations and are representative of values used popularly in decompression meters, dive tables, and dive planning software. Correlation functions were the model constrained permissble supersaturations which vary widely across dissolved gas and bubble models. Dissolved gas models admit greater supersaturations than bubble models.

Statistically, the outcomes of hit or no-hit were used as endpoints for this analysis. Other endpoints employed include Doppler bubble counts and various imaging metrics. All have their merits and the latter collect different information.

The USN and ZHL16 models correlate weakly:

$$\text{USN} - \chi^2 = 0.081$$
$$\text{ZHL16} - \chi^2 = 0.131$$

while the VPM and RGBM models correlate more strongly:

$$\text{VPM} - \chi^2 = 0.717$$
$$\text{RGBM} - \chi^2 = 0.861$$

Such might be expected in various quarters and this analysis thus both affirms and more importantly quantifies these speculations. While the correlation of VPM and RGBM with deep stop data is expected to be high, the correlation of USN and ZHL is low, but higher than might be expected intuitively. Perhaps the optimal approach to safe diving is a model somewhere between the extremes of each.

Any focus is safe operational diving. The LANL DB suggests that the diving profiles collected are low risk across the spectrum of activities (underlying incidence less than 1%), and the correlations of deep stop models (VPM and RGBM) within this DB also suggest that deep stop models are useful, practical, realistic, and reproducible for sport and technical diving with deep stop computers, software, and tables. Modern dive computer offerings with deep stop implementations are legion and increasing in number and this analysis supports their continued use for dive planning.

CORRELATED AND UNCORRELATED DIVEWARE

Diveware is mainly focused on a few staging algorithms, namely the Buhlmann ZHL, Work- man USN, Yount VPM, full and modified Wienke RGBM. The ZHL, USN, VPM and RGBM algorithms are validated by data, testing, and correlation analysis, seen herein for RGBM and VPM, and boast extensive usage in the technical diving community. The ZHL and USN algorithms are dissolved gas algorithms, while the RGBM and VPM are bubble algorithms. The former will yield shallow stop staging while the latter will result in deep stop staging. DCS spikes among users of all are not seen, suggesting rather safe staging protocols. User knobs allow conservative to liberal parameter settings depending on the vendor. In the dissolved gas algorithms, manipulation of critical tensions and gradients is the user knob. In the bubble algorithms, bubble size, permissible surfacing phase volume, and bubble gas diffusion lengths can be user manipulated.

Algorithms

For dissolved gas algorithms, the staging criteria is the usual $M-value$ approach, that is, for p the total tissue tension for all mixture gases (nitrogen and helium mainly) and M the critical tension for the particular tissue compartment, τ, we have,

$$p \leq M$$

always on ascent for all compartments. Stops are usually calculated for 10 *fsw* jumps upward with the longest wait across all compartments determining the stop time. Compartments range from $2 \leq \tau \leq 480$ *min* typically. Variations in M-values across software packages are small, usually in the so-called Spencer regime with M-values some 8 − 15% below the classical USN (Workman) values and close to the Buhlmann ZHL M-values. Reduction factors, ξ, as published by Wienke in fitting bubble models within M-value frameworks, can be applied to reduce M-values and reproduce deep stops, according to,

$$M = \xi G + P$$

with P ambient pressure and G the critical gradient computed from fixed values of critical tensions and ambient pressures. The reduction factors, ξ, are also called gradient factors, *GF*, in the technical diving community and are free floated in value (not constrained to bubble model correlations) for dive planning, with, roughly,

$$0.45 \leq GF \leq 0.95$$

as some measure to give deep stops to an otherwise shallow stop algorithm. Unlike the classical M-values and bubble factors ξ, (pseudo) GFs are not correlated with data in any form today. A variant of the above is called ratio deco, whereby,

$$\frac{p}{P} \leq R$$

and is just another representation of the same approaches above with ξs and GFs. For,

$$R = \frac{M}{P}$$

standard Haldane staging obtains. For reduction factors, ξ, in the modified RGBM,

$$R = \frac{\xi G + P}{P}$$

and in the GF scheme, with ζ some other set of free floating constants,

$$R = \frac{\zeta M}{P}$$

and all previous comments remain the same concerning ξ and ζ.

In bubble models like RGBM and VPM, a critical gradient, *G*, is limited by bubble volume and growth rates according to the separated phase volume, Φ,

$$\int_0^\infty dt \int_{r_0}^\infty \frac{\partial}{\partial t}(p - P)B(r)dr \leq \Phi$$

at all points and time, *t*, in the dive and at the surface, with *B* a functional representation of the excited bubble distribution and r_0 some critical excitation radius for growth. The distribution, *B*, and anywhere between 3 - 5 model parameters have been correlated in the RGBM to dive profile data and outcomes. Risk functions are also obtained from the data.

Packages

A potpourri of software packages available on the market are described briefly in the following. They are chosen because of their widespread use, utility, historical perspectives, and diver popularity. New ones are coming online too. They might be broadly categorized as dissolved gas, dissolved gas with GFs, pseudo-bubble, and bubble models. Dissolved gas, dissolved gas with GFs, and pseudo-bubble models are collectively also termed neo-Haldane models. In neo- Haldane models, M-values are reduced compared to the original USN, RN or Swiss values. The RGBM and VPM are the only true, full bubble models of interest and commercially available in diveware and computers.

- *ABYSS*

  ABYSS in the mid 90s first introduced the full RGBM into its diveware packages. The Buhlmann ZHL model was also included in the dissolved gas package. It has seen extensive use over the past 20 yrs or so in the technical diving area. A variety of user knobs on bubble parameters and M-values permit aggressive to conservative staging in both models. Both the ZHL and RGBM have been published and formally correlated with diving data. Later, the modified RGBM with ξs was incorporated into *ABY SS*. Modified RGBM with ξs was published and correlated with data in the late 90s and also served as the early basis for Suunto, Mares, Dacor, ConneXon, CressiSub, UTC, Mycenae, Aqwary, HydroSpace, and other RGBM dive computers. Full RGBM was first incorporated into HydroSpace computers, and today in Suunto, Atomic Aquatics, LiquiVision, and Steam Machines computers.

- *VPlanner*
  V Planner first introduced the VPM in the late 90s. Based on the original work of Yount and Hoffman, the software has seen extensive use by the technical diving community. No formal correlations of the VPM and *V Planner* with data have been published. User knobs allow adjustment of bubble parameters for aggressive to conservative staging. *V Planner* is also used in LiquiVision and Advanced Diving Corporation computers for technical diving.

- *ProPlanner*
  ProPlanner is a software package that uses modified M-values for diver staging. Buhlmann ZHL M-values with GFs are employed with user knobs for conservancy. The model is called the VGM by *ProP lanner* designers. Some GFs claim to mimic the VPM. Nothing has been formally published about the VGM and *ProP lanner* correlations.

- *GAP*
  GAP is a software package similar to *ABY SS* offering the full RGBM, modified RGBM with ξs, and Buhlmann ZHL with GFs. Introduced in the mid 90s, it has seen extensive usage in the recreational and technical sectors. Apart from user GFs, the models and parameters in *GAP* have been published and correlated with diving data and profiles tested over years. Adjustable conservancy settings for all models can be selected. *GAP* has been keyed to the Atomic Aquatics and LiquiVision dive computers.

- *DecoPlanner*
  DecoP lanner is a diveware package offered by the GUE training agency. Both the VPM and Buhlmann ZHL with GFs are available in *DecoP lanner*. Evolving over the past 10 - 15 yrs, *DecoP lanner* also incorporates GUE *ratio deco*, *p/P*, approaches to modifying GFs. *DecoP lanner* correlations with data have not been published. It has seen extensive use in the technical diving community.

- *RGBM Simulator*
  The *RGBM Simulator* is a software package marketed by HydroSpace Engineering and offers the full RGBM, modified RGBM with ξs, and Buhlmann

ZHL algorithms. All of the algorithms have been correlated with data and published. As with all other diveware, conservancy settings allow adjustments of bubble parameters, ξs, and M-values. On the market since the late 90s, *RGBM Simulator* is also keyed to the HydroSpace dive computer.

Output is typically extensive from modern diveware. Platforms range from PCs to Droid devices as well as mainframes. Languages employed in codes include VIZ, BASIC, FORTRAN, C, and derivatives. Meter vendors (Suunto, Mares, LiquiVision, UTC, Atomic, CressiSub to name a few) often supply their own software packages, keyed to the meter algorithms, for coupled dive planning.

Examples

Sample output [74] from 3 diveware packages is listed next. Example 1 is output from *ABY SS*, Example 2 is output from *GAP*, and Example 3 is output from *CCPlanner*. The profile entries are evident and the flow is easy to follow. In addition to decompression scheduling, packages also track oxygen toxicity. *CCPlanner* is a private LANL package.

Output from all diveware is similar to the offerings listed here. Essential features are always decompression schedules, ascent and descent rates, ambient surface pressure, CNS and full body toxicity estimates, gas switches and CCR setpoints, and user conservatism knob settings. The largest contingent of users are obviously technical divers, professional divers, scientific divers, and exploration divers across military, recreational, research, training, and commercial sectors.

CONSISTENT CRITICAL PARAMETER SETS

For Haldane computational algorithms, the process of constructing closed sets of time limits, tissue halftimes, and limiting tensions becomes an important activity. We detail a method for this closure, applying the approach to some exposure relationships. The approach maximizes the tissue perfusion equation, subject to a depth-time law (theoretical, fitted, inferred, or otherwise) at the exposure time limit, coupling exposure limits, halftimes, depths, and maximum tensions in the process.

As seen dissolved gas models limit tissue supersaturation, assuming that gas exchange is controlled by perfusion or diffusion in blood-tissue media. A perfusion equation quantifies bulk gas transfer,

$$\frac{\partial(p - p_a)}{\partial t} = -\lambda(p - p_a), \qquad \textbf{(311)}$$

ABYSS Output

ABYSS-2000, Advanced Dive Planning Software

This Short table printed for: Bruce Wienke on February 18, 2006

Profile: Abyss1 [Template=DEFAULT]

ATTENTION: The fact that this table was generated by ABYSS does not guarantee freedom from the possibility of decompression sickness. Diving is an inherently activity that may result in injury or death. Following this Abyss diving profile does not assure me that I won't be injured or killed. Decompression, Deep Diving, Ca Penetration and the use of Mixed Gas while diving are extremely hazardous aspects of an already dangerous activity

Surface Altitude, Safety Altitude, Algorithm,

J-Factors: **Depth =** **Bottom Time =** N2 = He = Ne = Ar =

Depth (Ft)	Time at	Run Time	Gas & Percent O2%	N2%	He%	Status
0	0.0	0.0	16	54	30	Surface
240	20.0	22.6	16	54	30	Entered by user
70	2.0	30.2	16	54	30	DECO
60	3.0	33.6	16	54	30	DECO
50	5.0	38.9	16	54	30	DECO
40	8.0	47.2	16	54	30	DECO
30	13.0	60.6	16	54	30	DECO
20	24.0	84.9	16	54	30	DECO
10	61.0	146.2	16	54	30	DECO
0	0.3	0.0	16	54	30	Surface

DECO Stops

Depth	Time	Run	Gas & Percent O2%	N2%	He%
70	2.0	30.2	16	54	30
60	3.0	33.6	16	54	30
50	5.0	38.9	16	54	30
40	8.0	47.2	16	54	30
30	13.0	60.6	16	54	30
20	24.0	84.9	16	54	30
10	61.0	146.2	16	54	30

DIVE SUMMARY

Run Time	146.6 Min	Deco Time	116.0 Min
CNS Clock	14.14%	OTU's	36.02
Max PPO2	1.32 (Atm)	Min PPO2	
Max END	158.08 (Ft)	Max Workload	Mild
RMV	0.50 (CuFt)	Max Depth	240.00 (Ft)

Gas Consumption

Gas & Percent O2%	N2%	He%	Volume (CuFt)	Reserve (CuFt)
16	54	30	257.0	0.0
			257.0	**0.0**

Total Gas Consumed	256.96
Required reserve	No reserve
Total Reserves	0.00
Total Gas Required	**256.96 (CuFt)**

GAP Output

NAUI GAP - Dive Information

Created for: brwswe | Page 1/1 | Creation Date 2/17/2006

The information presented here does in no way guarantee that you will not be injured or killed.
The authors accept no liability for your use of the information presented here.

deco schedule

NAUI-GAP (1530)

Model: RGBM classic RGBM

mode: OC

Depth	Time (RT)	Gas	PO2
30 m	22.0 (25)	Tx32/40	1.28
6 m	2 (29)	Tx32/40	0.51
3 m	6 (35)	Tx32/40	0.42
0 m	-- (35)	Tx32/40	0.00

Breathing gas	Volume
Tx32/40	2080.50 lit

Total CNS	Total OTU
13.5 %	35.8

Maximal Depth	Maximal Time
30.0 m	00:35 (= 35 min)

Maximal Po2 (B)	Maximal Po2 (D)
1.4 Bar	1.6 Bar

Maximal RMV (B)	Maximal RMV (D)
20.0 lit/min	10.0 lit/min

Maximal END
30.0 m

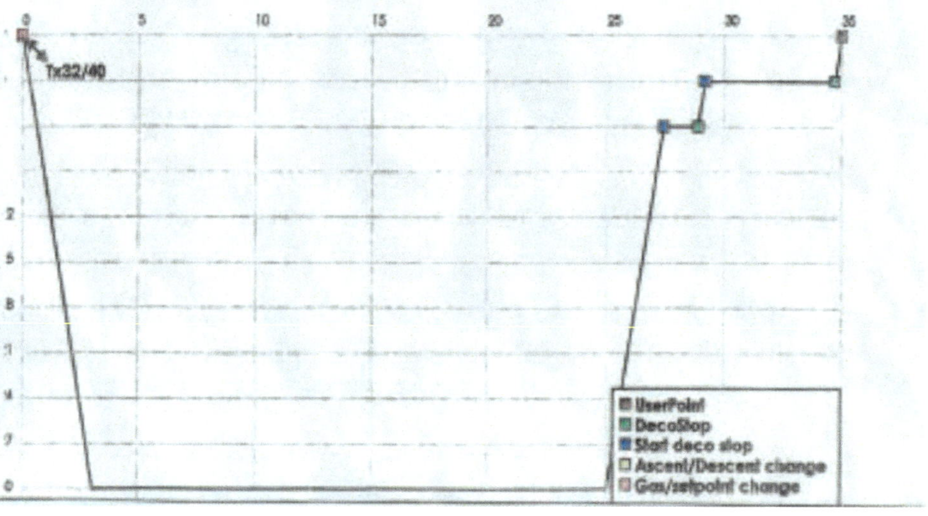

CCPlanner Output

```
                         CCPlanner Tmx 16-34

C&C Operations -- OC
Version 6/20/14

  --Dive Sequence Number--   1

  --Scaling/Control Flags--

  temperature = 303 K   repet flag =   1
  dive number =  1  altitude =      0.0 ft
  rfac = 0.85   pfac = 1.00   bfac = 0.35
  unsat =  1
  surface consumption rate =  0.75 ft^3/min
  tissue/m-value sets  jmix = 1 jndl = 1 nbub = 0
  phase constant =  840.0 microns^3
  zh = 0.61  zn = 0.81  trat = 1.03

  --Dive Profile--

  time since last dive =  1440. min
  time of last dive =     0. min
  ave depth last dive =      0. fsw

  surface air = 0.79 nitrogen  0.00 helium  down switches =  2

   switch 1 depth =     0.0 fsw helium = 0.00 nitrogen = 0.79
              speed =   60.0 fsw/min   way time =     0.0 min
   switch 2 depth =   190.0 fsw helium = 0.34 nitrogen = 0.50
              speed =  -30.0 fsw/min   way time =    17.5 min

  bottom trimix = 0.34 helium  0.50 nitrogen  up switches =  1

   switch 3 depth =    20.0 fsw helium = 0.00 nitrogen = 0.00
              speed =  -30.0 fsw/min   way time =     0.0 min

  fit parameters  rfit = 0.8589  bfit = 0.4697  pfit = 1.0000
                  rn = 0.9824  rh = 0.8596  r0 = 0.6718
                  depmx =  190.0  depav =  190.0

  ---Metrics And Debugging Parameters---

  rh =  0.8596 microns  rn =  0.9824 microns
  phase volume =    505. microns^3
  bfac/sfac/efac/tfac/dfac/gfac/rfac = 0.35/0.84/1.00/0.81/0.96/1.00/0.85/
  r0/rc/rstr/trat = 0.64/0.64/0.38/1.03/

  tissue  factor  tissue  factor
     1.0    1.00     0.3    1.00
     2.0    1.00     0.7    1.00
     5.0    1.00     1.7    1.00
    10.0    1.00     3.3    1.00
    20.0    1.00     6.7    1.00
    40.0    1.00    13.3    1.00
    80.0    1.00    26.7    1.00
   120.0    1.00    40.0    1.00
   160.0    1.00    53.3    1.00
   240.0    1.00    80.0    1.00
   320.0    1.00   106.7    1.00
   400.0    1.00   133.3    1.00
   480.0    1.00   160.0    1.00
   560.0    1.00   186.7    1.00
   720.0    1.00   240.0    1.00
```

CCPlanner Output

CCPlanner Tmx 16-34

trimix
fc = 1.00
alt = 0. fn = 0.50 fh = 0.34 fo = 0.16

	nitrogen		helium		
depth	ndl	half	ndl	half	
40.	190.3	240.0	190.3	80.0	0.0
50.	103.8	160.0	103.8	53.3	0.0
60.	54.0	80.0	54.0	26.7	0.0
70.	43.0	80.0	43.0	26.7	0.0
80.	36.0	80.0	36.0	26.7	0.0
90.	21.3	40.0	21.3	13.3	0.0
100.	18.4	40.0	18.4	13.3	0.0
110.	11.4	20.0	11.4	6.7	0.0
120.	10.0	20.0	10.0	6.7	0.0
130.	9.0	20.0	9.0	6.7	0.0
140.	8.2	20.0	8.2	6.7	0.0
150.	7.5	20.0	7.5	6.7	0.0
160.	5.2	10.0	5.2	3.3	0.0
170.	4.8	10.0	4.7	3.3	0.0
180.	4.4	10.0	4.4	3.3	0.0
190.	4.1	10.0	4.1	3.3	0.0
200.	3.8	10.0	3.8	3.3	0.0

m-values/mixed values

	nitrogen			helium		mixed	
half	m0	dm	half	m0	dm	mxm0	mxdm
1.0	119.8	2.12	0.3	153.3	2.59	133.4	2.31
2.0	115.0	2.04	0.7	147.4	2.48	128.1	2.22
5.0	100.5	1.79	1.7	129.8	2.17	112.4	1.94
10.0	81.7	1.47	3.3	103.5	1.70	90.5	1.56
20.0	68.3	1.27	6.7	83.8	1.40	74.6	1.32
40.0	60.2	1.19	13.3	72.0	1.28	65.0	1.22
80.0	53.9	1.12	26.7	63.0	1.18	57.6	1.14
120.0	51.1	1.10	40.0	59.3	1.14	54.4	1.12
160.0	49.4	1.08	53.3	57.3	1.12	52.6	1.10
240.0	47.1	1.06	80.0	55.5	1.11	50.5	1.08
320.0	45.8	1.05	106.7	55.0	1.10	49.5	1.07
400.0	44.9	1.05	133.3	54.7	1.09	48.9	1.07
480.0	44.2	1.04	160.0	54.5	1.09	48.4	1.06
560.0	43.7	1.04	186.7	54.4	1.08	48.0	1.05
720.0	42.7	1.02	240.0	54.1	1.07	47.3	1.04

tissue	tissue	radius	gradient
0.3	1.0	0.4503	103.2
0.7	2.0	0.4138	133.5
1.7	5.0	0.3828	163.9
3.3	10.0	0.3735	177.2
6.7	20.0	0.3681	184.6
13.3	40.0	0.3652	188.5
26.7	80.0	0.3636	190.5
40.0	120.0	0.3631	191.1
53.3	160.0	0.3628	191.5
80.0	240.0	0.3626	191.8
106.7	320.0	0.3624	192.0
133.3	400.0	0.3623	192.1
160.0	480.0	0.3623	192.2
186.7	560.0	0.3622	192.2
240.0	720.0	0.3622	192.3

CCPlanner Output

CCPlanner Tmx 16-34

trimix
fc = 1.00
alt = 0. fn = 0.50 fh = 0.34 fo = 0.16

	nitrogen		helium		
depth	ndl	half	ndl	half	
40.	190.3	240.0	190.3	80.0	0.0
50.	103.8	160.0	103.8	53.3	0.0
60.	54.0	80.0	54.0	26.7	0.0
70.	43.0	80.0	43.0	26.7	0.0
80.	36.0	80.0	36.0	26.7	0.0
90.	21.3	40.0	21.3	13.3	0.0
100.	18.4	40.0	18.4	13.3	0.0
110.	11.4	20.0	11.4	6.7	0.0
120.	10.0	20.0	10.0	6.7	0.0
130.	9.0	20.0	9.0	6.7	0.0
140.	8.2	20.0	8.2	6.7	0.0
150.	7.5	20.0	7.5	6.7	0.0
160.	5.2	10.0	5.2	3.3	0.0
170.	4.8	10.0	4.7	3.3	0.0
180.	4.4	10.0	4.4	3.3	0.0
190.	4.1	10.0	4.1	3.3	0.0
200.	3.8	10.0	3.8	3.3	0.0

m-values/mixed values

	nitrogen			helium		mixed	
half	m0	dm	half	m0	dm	mxm0	mxdm
1.0	119.8	2.12	0.3	153.3	2.59	133.4	2.31
2.0	115.0	2.04	0.7	147.4	2.48	128.1	2.22
5.0	100.5	1.79	1.7	129.8	2.17	112.4	1.94
10.0	81.7	1.47	3.3	103.5	1.70	90.5	1.56
20.0	68.3	1.27	6.7	83.8	1.40	74.6	1.32
40.0	60.2	1.19	13.3	72.0	1.28	65.0	1.22
80.0	53.9	1.12	26.7	63.0	1.18	57.6	1.14
120.0	51.1	1.10	40.0	59.3	1.14	54.4	1.12
160.0	49.4	1.08	53.3	57.3	1.12	52.6	1.10
240.0	47.1	1.06	80.0	55.5	1.11	50.5	1.08
320.0	45.8	1.05	106.7	55.0	1.10	49.5	1.07
400.0	44.9	1.05	133.3	54.7	1.09	48.9	1.07
480.0	44.2	1.04	160.0	54.5	1.09	48.4	1.06
560.0	43.7	1.04	186.7	54.4	1.08	48.0	1.05
720.0	42.7	1.02	240.0	54.1	1.07	47.3	1.04

tissue	tissue	radius	gradient
0.3	1.0	0.4503	103.2
0.7	2.0	0.4138	133.5
1.7	5.0	0.3828	163.9
3.3	10.0	0.3735	177.2
6.7	20.0	0.3681	184.6
13.3	40.0	0.3652	188.5
26.7	80.0	0.3636	190.5
40.0	120.0	0.3631	191.1
53.3	160.0	0.3628	191.5
80.0	240.0	0.3626	191.8
106.7	320.0	0.3624	192.0
133.3	400.0	0.3623	192.1
160.0	480.0	0.3623	192.2
186.7	560.0	0.3622	192.2
240.0	720.0	0.3622	192.3

with the exchange of inert gas driven by the local gradient, that is, the difference between arterial blood, p_a, and local tissue tension, p. Obviously the exchange process is very complicated, and models are only approximate. The solutions are well known, simple classes of exponential functions, bounded by arterial and initial tissue tensions, p_a and p_i,

$$(p - p_a) = (p_i - p_a)\ exp\ (-\lambda t), \tag{312}$$

with λ the decay rate, defined in terms of the halftime, τ ,

$$\lambda = \frac{.693}{\tau} \tag{313}$$

with instantaneous tissue tension, p, in that compartment. Compartments with 2, 5, 10, 20, 40, 80, 120, 240, 360, 480, and 720 minute halftimes, τ, are employed in applications, and halftimes are assumed to be independent of pressure.

Next, algorithms limit degrees of dissolved gas buildup, p, hypothetical absolute compartment supersaturation, by *critical* values, M, such that,

$$p \leq M, \tag{314}$$

across all compartments at all times during exposure, and upon surfacing. Equivalently, critical ratios, R, and critical gradients, G are also employed, with,

$$R = \frac{M}{P}, \tag{315}$$

$$G = M \quad P, \tag{316}$$

for ambient pressure, P. Critical parameters evolved from self consistent application of assumed tissue response functions to sets of exposure data, that is, trial and error bootstrapping of model equations to observed exposure time limits. Newer compilations ultimately extend older ones to extended data ranges.

In a diffusion framework, nonstop air limits, t_n, roughly satisfy a bulk transfer law,

$$dt_n^{1/2} = 465 \ fsw \ min^{1/2}, \qquad \textbf{(317)}$$

at depth, *d* (Hempleman *square root* law), generalized by writing,

$$dt^a = b, \qquad \textbf{(318)}$$

for *a* and *b* some constants. Ranges subtended today in tables and meters include:

$$0.25 \leq a \leq 0.65, \qquad \textbf{(319)}$$

$$250 \ fsw \ min^a \leq b \leq 500 \ fsw \ min^a, \qquad \textbf{(320)}$$

A separated phase model for nonstop air diving suggests,

$$\delta \ d \ (t_n + 1/\lambda) = 8750 \ fsw \ min, \qquad \textbf{(321)}$$

for number factor, δ, collectively representing bubble seeds excited by compression-decompression, that is, from surface pressure, P_0, to ambient pressure, P, and back to P_0,

$$\delta = \frac{P}{P_0} - 1. \qquad \textbf{(322)}$$

The phase law generalizes obviously to,

$$\delta \ d \ (t + 1/\lambda)^a = \frac{d^2}{P_0 + d} \ (t + 1/\lambda)^a = b, \qquad \textbf{(323)}$$

with,

$$P = P_0 + d. \qquad \textbf{(324)}$$

The depth-time law and tissue equation present a minimax problem, here, maximization of the tissue equation at depth subject to the constraint of the depth-time law. The standard approach sets the depth derivative of the tissue equation (tension) to zero at the exposure time limit, under the primary constraint of the depth-time equation. First writing ambient gas partial pressure, p_a, as

$$p_a = p_0 + fd, \quad \textbf{(325)}$$

for surface partial pressure, p_0, mole fraction, f, and then differentiating tension, p, with respect to depth, d, we find in general,

$$\frac{\partial p}{\partial d} = f - f\ exp\ (-\lambda t) + (p_0 + fd - p_i)\lambda\ exp\ (-\lambda t)\frac{\partial t}{\partial d} = 0, \quad \textbf{(326)}$$

as the maximization condition. The time derivative with respect to depth, $\partial t/\partial d$, is evaluated from the assumed exposure law (theoretical, fitted, inferred), and then inserted above. The resulting expression couples halftime, τ, to exposure limit, t_n, and the value of the tissue tension at those values is the (maximized) critical tension, M_0.

Some table and meter algorithms still rely on (Haldane) dissolved gas treatments to schedule diving, with square root-like nonstop limits folded into a multitissue perfusion framework. So, as example, consider the bulk relationship, so that,

$$\frac{\partial t}{\partial d} = -\frac{t}{ad}. \quad \textbf{(327)}$$

Setting $p_i = p_0$, substituting the derivative, and maximizing at the nonstop limit, t_n, there results,

$$1 - exp\ (-\lambda t_n) - \frac{\lambda t_n}{a}\ exp\ (-\lambda t_n) = 0. \quad \textbf{(328)}$$

At nonstop time, $t = t_n$, the tissue tension is maximized, that is, $p = M_0$, so that,

$$M_0 = p_a + (p_i - p_a)\frac{a}{a + \lambda t_n}. \quad \textbf{(329)}$$

The maximization condition links λ and t_n together, while M_0 falls out of the tissue equation. The quantity λt_n is pivotal to the solution. Table **4d** gives thumbnail solutions to,

$$exp\ (x) = 1 + \frac{x}{a}, \quad \textbf{(330)}$$

for a, as function of dimensionless parameter, $x = \lambda t_n$.,

Table 4d. Maximization parameters.

a	1/*a*	*x*
.157	6.37	3.00
.323	3.09	2.00
.435	2.29	1.50
.455	2.20	1.40
.488	2.05	1.30
.500	2.00	1.25
.517	1.93	1.20
.549	1.82	1.10
.581	1.72	1.00
.771	1.29	0.50
.951	1.05	0.10
1.00	1.00	0.00

In the separated phase model, we have differentiating,

$$\frac{\partial t}{\partial d} = -\frac{2(t + 1/\lambda)}{ad}, \quad \textbf{(331)}$$

so that,

$$1 - exp\ (-\lambda t_n) - \frac{2(\lambda t_n + 1)}{a}\ exp\ (-\lambda t_n) = 0, \qquad \textbf{(332)}$$

as the maximization constraint. Or, equivalently, one needs,

$$exp\ (y) = 1 + \frac{2(y+1)}{a}, \qquad \textbf{(333)}$$

with $y = \lambda t_n$. Then, the critical tension, M_0, is given by,

$$M_0 = p_a + (p_i - p_a)\ \frac{a}{a + 2(\lambda t_n + 1)} \qquad \textbf{(334)}$$

The procedures for constructing a consistent set can be summarized as follows;

- first, from experiment, wet or dry tests, Doppler, or otherwise, a set of nonstop time limits, t_n, at depth, d, is obtained;
- next, the set is fitted to the two parameter power law given above, and the constants a and b determined;
- then, with a and b determined, the controlling halftime, τ, is obtained from t_n at d;
- finally, from τ, t_n, d, and a, the critical tension, M_0, is extracted, closing the whole set;
- compute and recalibrate parameters against a set of test profiles, data, or exposure information.

The set a, b, τ, d, t_n, x, y, and M_0 then close self consistently when derived according to the above set of equations and constraints.

Results for some air limits are summarized in Table **4e**, using a standard nonlinear least squares (NLS) approach in fitting t_n to the depth-time relationship with L_2 error norm, the square root of the sum of the squares of the differences in the fit. Labels RGBM, DCIEM, ZHL, Spencer, and USN refer to nonstop time limits [15, 51, 74, 85] for the popular models.

Table 4e. Fits, limits, halftimes, and critical tensions.

	RGBM	DCIEM	ZHL	Spencer	USN
Fit Parameters					
a	.94	.48	.46	.39	.41
$b(fsw\ min^a)$	6119	362	385	290	355
x	1.40	1.34	1.39	1.65	1.58
y	2.00	1.08	1.65	1.16	1.57
$L_2(fsw)$	12.8	57.7	62.3	85.5	56.5
Nonstop Limits $t_n(min)$					
$d(fsw)$					
30	200	150	290	225	
40	110	90	125	135	200
50	70	70	75	75	100
60	50	50	54	50	60
70	35	35	38	40	50
80	26	25	26	30	40
90	20	20	20	25	30
100	16	15	20	20	25
110	13	12	17	15	20
120	11	10	15	10	15
130	9	8	11	5	10
Halftimes τ (min)/Critical Tensions $M_0(fsw)$					
$d(fsw)$					
30	69/46	98/44	122/45	134/45	178/45
40	38/53	53/49	68/51	60/52	88/52
50	24/60	33/55	42/57	37/58	52/57
60	17/67	22/61	28/63	23/64	33/64
70	12/74	17/67	20/69	16/71	23/69
80	9/79	12/73	15/73	11/77	16/76
90	7/87	10/87	12/82	8/84	12/83
100	6/94	8/85	9/88	7/90	10/89
110	5/101	6/91	8/84	6/96	8/95
120	4/108	5/96	6/100	4/103	6/101
130	3/114	4/102	5/106	3/109	5/107

The structure and range above is interesting. Surfacing critical tensions, M_0, and tissue half- times, τ, are bounded,

$$44\ fsw \leq M_0 \leq 114\ fsw \quad \textbf{(335)}$$

$$3\ min \leq \tau \leq 178\ min \quad \textbf{(336)}$$

with the DCIEM, ZHL, Spencer, and USN (all Haldane models) exhibiting roughly similar parameter clustering, but with the RGBM (phase model) rather different from the rest. And all said with nonstop time limits, t_n, pretty much the same across all models. In diving practice, this is just another manifestation of differences between dissolved gas and phase models. As seen, deeper stops and overall shorter decompression times are the result. And that brings up some important points for the following.

SHALLOW AND DEEP STOP DATA CORRELATION

We purposely correlated deep stop data with deep stop models previously because this is an important aspect of diving validation that hasn't been completed in the past. The question of deep stops *versus* shallow stops is one of current interest. Players often fall into one category or another religiously. Deep stops are relatively new on the diving scene and somewhat untested, while shallow stops have been around for almost century, are tested, correlated with data, and also used routinely. The growing body of diving experience using deep stop tables, meters, and software attests to the viability of deep stops in diving regimens. By that is meant that DCS spikes using the new bubble models are not seen in diving populations across many disciplines, and no more so than earlier models [66]. Anecdotal information released by meter vendors suggests the same [75]. This is not science, of course, but noteworthy across the whole of diving activity and staging protocols. On more scientific note, one study [74, 92] of dive computers usage across 1000s of dives showed nearly identical (low) DCS incidence rates for both RGBM and ZHL decompression meters, around 1.4%. So DCS appears neither singular nor prevalent in real user communities for both models.

One fact bears here. From purely physical dynamics, there are two ways to effect bubble and dissolved gas elimination in diver staging. Both work, but for different reasons. Shallow stops focus on dissolved gas elimination while deep stops focus on minimizing bubble growth. Both must treat bubbles in all zones, of course, but from experience, and profile statistics, deep stops appear more efficient and result in shorter decompression time. Physiologically, the jury is still out.

CHAPTER 3

Applications

Abstract: Model applications are important as predictors of comparative diver staging and protocols. Whether by computer, diveware, or tables, planning is requisite for safe diving. Applications following focus on both shallow and deep mixed gas diving on OC and RB systems. Profiles along with estimated risk are tallied. A bubble model correlated with deep stop diver outcomes is used. Profiles and schedules are listed and compared. Many profiles reside as data entries in the LANL DB. So the applications are both data and comparative dive profiles.

Keywords: Comparative gas transport, Real profiles, Risk, Staging.

Applications with model and risk estimates follow. Applications analyses [75, 79] include the Marroni and Bennett 2.5 *min* recreational deep stop, the C & C 450/20 multiple RB dive sequence at 1.4 *stm*, NEDU deep stop tests, French Navy deep stop profiles, EXPLORER decompression meter algorithm, NAUI Tables, University of Wisconsin Seafood Diver Tables, comparative NAUI, PADI, Oceanic NDLs and repetitive dives, comparative nitrogen and helium mixed gas risks, USS Perry deep RB exploration dive, world record OC dive, and WKPP extreme cave exploration profiles. The LANL model enjoys useful, widespread, and prudent application in mixed gas diving, both in recreational and technical sectors, and forms the bases of software, released tables and decompression meters used by scientific, commercial, and re- search divers. Supercomputing power is employed for application and correlation of model and data.

NONSTOP AND REPETITIVE AIR DIVING

Nonstop limits (NDLs), denoted t_{nn}, from the US Navy, PADI, NAUI, and ZHL (Buhlmann) (Tables **15, 16**) provide a set for comparison of relative DCS risk.

B.R. Wienke

Listed in Table **5a** are the NDLs and corresponding risks for the nonstop excursion, assuming ascent and descent rates of 60 *fsw/min* (no safety nor deep stops). Dissolved gas and phase risk estimates vary little for cases, and only the phase estimates are included. Surface intervals (SIs) between dives are time spent at the surface.

Table 5a. Risk estimates for standard air NDLs.

	USN NDL	Risk	PADI NDL	Risk	NAUI NDL	Risk	ZHL NDL	Risk
d (*fsw*)	t_n (*min*)	β	t_n (*min*)	β	t_n (*min*)	β	t_n (*min*)	β
35	310	4.3%	205	2.0%			181	1.3%
40	200	3.1%	140	1.5%	130	1.4%	137	1.5%
50	100	2.1%	80	1.1%	80	1.1%	80	1.1%
60	60	1.7%	55	1.4%	55	1.4%	57	1.5%
70	50	2.0%	40	1.2%	45	1.3%	40	1.2%
80	40	2.1%	30	1.3%	35	1.5%	30	1.3%
90	30	2.1%	25	1.5%	25	1.5%	24	1.4%
100	25	2.1%	20	1.3%	22	1.4%	19	1.2%
110	20	2.2%	13	1.1%	15	1.2%	16	1.3%
120	15	2.0%	13	1.3%	12	1.2%	13	1.3%
130	10	1.7%	10	1.7%	8	1.3%	10	1.7%

Risks are internally consistent across NDLs at each depth, and agree with the US Navy assessments in Table **4b**. Greatest underlying risks occur in the USN shallow exposures. The PADI, NAUI, and ZHL risks are all less than 2% for this set, and risks for single DCS incidence are less than 0.02. PADI and NAUI have reported that incidence rates (*p*) across all exposures are less than 0.001%, so considering their enviable track record of diving safety, our estimates are liberal. ZHL risk estimates track as the PADI and NAUI risks, again, very safely. Estimates were corroborated [74] within data sets at Duke both in Tables **5a** and **5b**.

Next, the analysis is extended to profiles with varying ascent and descent rates, safety stops, and repetitive sequence [53, 78 - 80, 82]. Table **5b** lists nominal profiles (recreational) for various depths, exposure and travel times, and safety stops at 5 *msw*. Mean DCS estimates, r, are tabulated for both dissolved gas supersaturation ratio and excited bubble volume risk functions, with nominal variance, $r_{\pm} = r \pm 0, 004$, across all profiles.

Table 5b. Dissolved and separated phase risk estimates for nominal profiles.

Profile (*depth/time*)	Descent Rate (*msw/min*)	Ascent Rate (*msw/min*)	Safety Stop (*depth/time*)	Risk β	Risk σ
14 *msw*/38 *min*	18	9	5 *msw*/3 *min*	0.0034	0.0062
19 *msw*/38 *min*	18	9	5 *msw*/3 *min*	0.0095	0.0110
28 *msw*/32 *min*	18	9		0.0200	0.0213
37 *msw*/17 *min*	18	9	5 *msw*/3 *min*	0.0165	0.0151
18 *msw*/31 *min*	18	9	5 *msw*/3 *min*	0.0063	0.0072
	18	9		0.0088	0.0084
	18	18		0.0101	0.0135
	18	18	5 *msw*/3 *min*	0.0069	0.0084
17 *msw*/32 *min* SI 176 *min*	18	9	5 *msw*/3 *min*		
13 *msw*/37 *min* SI 176 *min*	18	9	5 *msw*/3 *min*		
23 *msw*/17 *min*	18	18	5 *msw*/3 *min*	0.0127	0.0232

The ZHL (Buhlmann) NDLs and staging regimens are widespread across decompression meters presently, and are good representations for dissolved gas risk analysis. The RGBM is newer, more modern, and is coming online in decometers and associated software. For recreational exposures, the RGBM collapses to a dissolved gas algorithm. This is reflected in the risk estimates above, where estimates for both models differ little [9, 20, 23, 36, 86].

Simple comments hold for the analyzed profile risks. The maximum relative risk is 0.0232 for the 3 dive repetitive sequence according to the dissolved risk estimator. This translates to 2% profile risk, which is comparable to the maximum NDL risk for the PADI, NAUI, and ZHL NDLs. This type of dive profile is common, practiced daily on liveaboards, and benign. According to Gilliam, the absolute incidence rate [93] for this type of diving is less than 0.02%. Again, our analyses overestimate risk. Effects of slower ascent rates and safety stops are seen only at the 0.25% to 0.5% level in relative surfacing risk. Safety stops at 5 *msw* for 3 *min* lower relative risk an average of 0.3%, while reducing the ascent rate from 18 *msw/min* to 9 *msw/min* reduces relative risk an average of 0.35%. Staging, NDLs, and constraints imposed by decometer algorithms are consistent with acceptable and safe recreational diving protocols. Estimated absolute risk associated across all ZHL NDLs and staging regimens analyzed herein is less than 2.32%, probably much less in actual practice. That is, we use $p = 0.0069$, and much evidence suggests $p < 0.0001$, some ten times safer.

Implicit in such formulations of risk tables are assumptions that given decompression stress is more likely to produce symptoms if it is sustained in time, and that large numbers of separate events may culminate in the same probability after time integration. Though individual schedule segments may not be replicated enough to offer total statistical validation, categories of predicted safety might be grouped within subsets of corroborating data. For instance, risks on air dives might be estimated from just nitrox data, risks on trimix from just trimix data, risks on heliox just from heliox data, *etc.* Since the method is general, any model parameter or meaningful index, properly defined, can be applied to decompression data, and the full power of statistical methods employed to quantify overall risk. While powerful, such statistical methods are neither deterministic nor mechanistic, and cannot predict on first principles. But as a means to table fabrication with quoted risk, such approaches offer attractive pathways for analysis.

Questions of what risk is acceptable to the diver vary. Sport and research divers would probably opt for small risk (1% or less), while military and commercial divers might live with higher risk (5%), considering the nearness of medical attention in general. Many factors influence these two populations, but fitness and acclimatization would probably play strategically.

RECENT DOPPLER AND WET TEST ANALYSES

Recent Doppler and wet tests are interesting, including our recorded CCR 16 dive sequence to 450 *fsw*. Gas transport [13, 84] analysis of these applications follows, along with bubble risk estimates.

BENNETT AND MARONNI 2.5 MINUTE RECREATIONAL DEEP STOP

Deep stops are already mainliners in some training agency protocols for no and light de- compression diving on air and nitrox. The prescription is to make a deep stop at half depth for 1 - 3 *min*, followed by a shallow stop in the 15 *fsw* zone for 1 - 2 *min*. In Table **6a**, we cite bubble surfacing risks for a deep stop at half depth for 1 *min*, 2.5 *min*, and 4 *min*, the middle case suggested by Bennett and Maronni from Doppler scoring [74, 94], followed by direct ascent to the surface. Surfacing supersaturation risks are tabulated in Table **6b** for comparison. Dives are carried out to the (old) US Navy NDLs for easy reference. Deep stops for less than 2.5 *min* reduce recreational risk out to the Navy NDLs in all cases. Bubble risks decrease for short deep stops and then increase as stop times increase. As stop times continue to increase, the dives will require decompression. In other words, with increasing deep stop time, the dives become multilevel decompression dives. Obviously, the playoff of deep stop time against bot- tom time is a minimax problem. This is traced to bubble behavior with increased gas tensions for increasing deep stop time. In all cases, stop time in the shallow zone was 1 *min*. Longer stop times in the shallow zone had little effect on surfacing risks. Shallow stops in training regimens probably serve better to teach buoyancy control to neophytes.

Table 6a. Comparative bubble risks for recreational deep stops.

Depth (*fsw*)	Time (*min*)	No Stop β	1 *min* Stop β	2.5 *min* Stop β	4 *min* Stop β
80	40	2.10%	1.93%	1.90%	1.91%
90	30	2.10%	1.87%	1.83%	1.84%
100	25	2.10%	1.74%	1.71%	1.72%
110	20	2.20%	1.65%	1.61%	1.62%
120	15	2.00%	1.50%	1.46%	1.47%
130	10	1.70%	1.29%	1.25%	1.26%

Ascent and descent rates were standard in the analysis, that is, 30 *fsw/min* and 60 *fsw/min* respectively. The small risk spread for 1 - 4 *min* accommodates recreational deep stop training regimens, that is, 1 - 3 *min* deep half stop for many agencies.

Corresponding supersaturation risks in Table **6b** are seen to increase monotonically with length of deep stop. This is to be expected in dissolved gas models, with exposures at increasing depths for increasing times cascading tissue tensions, oblivious to any bubble-dissolved gas interactions tracked in Table **6a**.

Table 6b. Comparative supersaturation risks for recreational deep stops.

Depth (*fsw*)	Time (*min*)	No Stop σ	1 *min* Stop σ	2.5 *min* Stop σ	4 *min* Stop σ
80	40	2.10%	2.12%	2.18%	2.26%
90	30	2.10%	2.13%	2.20%	2.29%
100	25	2.10%	2.15%	2.23%	2.34%
110	20	2.20%	2.24%	2.32%	2.41%
120	15	2.00%	2.10%	2.20%	2.38%
130	10	1.70%	1.78%	1.91%	2.13%

C & C TEAM 450/20 MULTIPLE RB DIVE SEQUENCE AT 1.4 *ATM*

Details of a 16 dive sequence by members of the C & C Team to 450 *fsw* for 20 *min* at 1.4 *atm* follow. Dives were successfully completed in tandem without mishap, and are included in the LANL Data Bank. All dives follow the same schedule, as given in Table **7**. Oxtox (both CNS and full body) metrics are included. Diver Tags and Outcomes are recorded according to the LANL Data Bank profile schema described previously. Diver Tag 1 is one of the authors (BRW). Risk estimates (both bubble and supersaturation) are noted, along with binomial probabilities for 16 tandem dives within a LANL Data Bank underlying incidence rate of 0.69%. Diluent is 10/80 trimix with a pp_{O2} setpoint of 1.4 *atm*. The cumulative CNS clock fractions exceed a limit of 1.0, while OTU uptake remains below a limit of 650 *min*. There is likely greater variability in oxtox limit points than decompression limit points. Descent and ascent rates are standard, except in the 30 *fsw* zone where the ascent rate is 1 *fsw/min*.

Zero hit probability is *P* (0), while probability of one hit is *P* 1. The probability of two or more hits is vanishingly small. Computed bubble risk, β, is below binomial probability, *P* (1).

Table 7. RB 16 dive sequence at 1.4 *atm*.

Depth (*fsw*)	Time (*min*)	CNS Clock (*fraction*)	OTU Uptake (*min*)
450	20	.17	32.6
360	0.5	.01	0.8
350	0.5	.01	0.8
340	0.5	.01	0.8
330	0.5	.01	0.8
320	0.5	.01	0.8
310	0.5	.01	0.8
300	1.0	.02	1.6
290	1.0	.02	1.6
280	1.0	.02	1.6
270	1.0	.02	1.6
260	1.0	.02	1.6
250	1.0	.02	1.6
240	1.0	.02	1.6
230	1.5	.03	1.8
220	1.5	.03	1.8
210	2.0	.03	4.1
200	2.0	.03	4.1
190	2.0	.03	4.1
180	2.0	.03	4.0
170	2.0	.02	4.0
160	2.5	.02	4.0
150	2.5	.02	3.9
140	3.5	.03	5.7
130	5.0	.05	9.0
120	5.0	.04	8.5
110	5.0	.04	8.4
100	5.5	.04	9.0
90	6.0	.05	9.8
80	8.0	.07	13.0
70	8.0	.07	12.5

(Table 7) contd.....

Depth (*fsw*)	**Time** (*min*)	**CNS Clock** (*fraction*)	**OTU Uptake** (*min*)
60	9.5	.08	15.5
50	11.0	.10	17.9
40	12.0	.10	19.5
30	8.5	.07	13.8
20	10.5	.09	17.1
10	17.0	.11	25.2
	211.5	1.38	262.2

Dive Tags = 2042 - 2058

Diver Tags = 3,20,5,**1**,9,6,10,2,14,4,15,7,8,11,16,12

Diver Outcomes = 3,4,3,**3**,4,3,4,3,4,3,3,3,4,3,4,3

Underlying Incidence = 20/2879

$\beta = 4.27\%$, $\sigma = 12.67\%$

$P(0) = 89.4\%$, $P(1) = 10.4\%$

NEDU DEEP STOP AIR TESTS

The Navy Experimental Diving Unit recently tested their version of air deep stops [74, 95] with a moderate DCS rate. Profiles tested are given in Table **8**, along with a suggested LANL deep stop profile. Profile NEDU 1 incurred a 5.5% DCS hit rate, while NEDU 2 1 incurred a lower 1.5% DCS hit rate.

Table 8. Comparative NEDU air deep stop schedules.

	NEDU 1	**NEDU 2**	**LANL**
Depth (*fsw*)	**Time** (*min*)	**Time** (*min*)	**Time** (*min*)
170	30	30	30
120			0.5
110			1.5
100			2.5
90			3.5
80			4.5
70			5.0
70	12		5.0

(Table 8) contd.....

	NEDU 1	NEDU 2	LANL
Depth *(fsw)*	**Time** *(min)*	**Time** *(min)*	**Time** *(min)*
60	17		7.0
50	15		11.0
40	18	9	14.5
30	23	23	22.0
20	17	52	28.5
10	72	93	59.9
	206	207	195
σ	5.6%	2.4%	3.4%
β	10.6%	3.2%	2.6%

Bubble risk is higher in both NEDU 1 and NEDU 2, but large in NEDU 1. NEDU 1 is a multilevel decompression dive with inadequate treatment in the shallow zone. Initial deep stops in NEDU 1 did not control bubble growth, and the length of the stay in 70, 60, and 50 *fsw* builds up dissolved gas in the middle range tissues, which then diffuses into bubbles causing them to grow. NEDU 2 is classic with no deep stops, and very long times in the shallow zone to effect decompression. The LANL schedule has deeper stops, shorter midzone times, and then shorter times in the shallow zone compared to both NEDU 1 and NEDU 2. One important factor here is the shape of the decompression schedule, that is the LANL profile is shorter overall, with NEDU 1 and NEDU 2 profiles exhibiting supersaturation staging with shallow belly and tail, while the LANL profile is steeper exhibiting bubble staging with deeper stops and steeper ascent rate. Both NEDU profiles are not of the genre typically dived by users of modern deep stop tables, software, and meters.

Gas transport [13, 84] analyses on both NEDU schedules suggests that NEDU 1 produces 15% - 30% larger bubble volumes on surfacing, due to the longer stay in the mid zone, while NEDU 2 produces surfacing bubble volumes 3% - 5% larger than surfacing bubble volumes in the LANL profile. Surfacing bubble volumes in the LANL profile were close to the staging limit point.

FRENCH NAVY DEEP STOP SCHEDULES

The French Navy also tested deep stop air schedules [74, 96]. Three protocols on deep air were employed and none exhibited Grade 4 Doppler bubbles. Analysis centered on just Grade 3 bubbles. For purposes of deep stop analysis, Protocol 1, a dive similar to NEDU 1, is interesting. Protocol 1 is a deep air dive to 200 *fsw* for 20 *min*, with ascent staging according to Table **9**. Contrasting staging strategies are denoted MN90, the standard French Navy dissolved gas regimen, and LANL. Outside of World Navies, few diving sectors today even contemplate air decompression diving to 200 *fsw*. Risks in air dives beyond 150 *fsw* are reported to increase by factors of 10 over similar dives at shallower depth [65, 86]. This is, of course, one major reason why trimix and heliox become mixtures of choice for deep and decompression diving worldwide, across commercial, scientific, exploration, and research sectors.

Table 9. French Navy air deep stop schedules at 200 *fsw*.

	Protocol 1	MN90	LANL
Ascent Rate *fsw/min* Starting at 90 *fsw* Depth (*fsw*)	**10 Time (*min*)**	**20 Time (*min*)**	**30 Time (*min*)**
200	20	20	20
130			0.5
120			0.5
110			1.0
100			1.0
90			1.0
80	1		1.5
70	1		2.0
60	2		2.0
50	2		2.5
40	4		3.0
30	6	3	6.0
20	9	8	7.0
10	22	32	8.0

(Table 9) contd.....

	Protocol 1	**MN90**	**LANL**
Ascent Rate *fsw/min* Starting at 90 *fsw* Depth (*fsw*)	**10 Time (*min*)**	**20 Time (*min*)**	**30 Time (*min*)**
	78	68	65
β	3.9%	2.2%	2.1%

By contrast, LANL staging starts deeper, is shorter overall, and has smaller bubble risk than Protocol 1. Protocol 1, however, tracks more closely with LANL than NEDU 1, and exhibits lower risk than NEDU 1. However, run time for Protocol 1 *versus* MN90 is longer, unlike conventional bubble model run times. Estimated bubble risks, β, are tabulated at the bottom of Table **9**.

GAS TRANSPORT ANALYSIS

With regard to the preceeding dives and schedules, a couple of points are interesting. These follow from a closer look at dissolved and bubble gas phases across the profiles, using LANL tools and selected way points on the dives. These comments also apply to deep and decompression staging using traditional dissolved gas models and tables [13, 78, 89]. Remember these comments are made within the LANL model framework and attendant data correlation:

- bubble growth in the deep zone of decompression profiles NEDU 1 and Protocol 1 is not constrained in their version of deep stop air tests;
- deep stops are not deep enough in NEDU 1 and Protocol 1, nor are follow stops;
- critical phase volume limit points are exceeded in NEDU 1 and Protocol 1 even before the diver exits, in other words, along the decompression glide path underwater;
- the recreational 2.5 *minute* stop at 1/2 depth within the NDLs of even the old USN tables maintains the phase volumes below limit points;
- the LANL 450/20 profiles also surface below the phase volume limit point, no surprise because profiles were designed to meet that constraint;
- supersaturation profiles MN90 and NEDU 2 also do not control bubble growth in the deeper zones, but the separated phase volume is below model limit points, with pressure in the shallow zone sufficient to constrain bubble growth and

maintain adequate dissolution, but time consuming because bubbles are now larger in the shallow zone.

Much the same can be said of supersaturation *versus* bubble staging strategies in general.

TABLE, METER, AND PROFILE RISK ANALYSES

To finish up analysis, consider other applications across tables, meters, and software, focusing on shallow stop *versus* deep stop profiles, risk, and data. The time span of these applications is the past 3 - 5 *yrs*, and they represent real mixed gas diving across many venues.

UW SEAFOOD DIVER AIR TABLES

As another application of the LANL Data Bank to table construction and analysis, we detail a set of tables of interest to the University of Wisconsin (UW), along with estimated risk for various nonstop limits gleaned from the data. These Tables have no groups, and simple rules. Released mixed gas RGBM Tables resulted from similar analyses across both the technical and recreational segments. Such Tables are certainly useful for a broad spectrum of diving, and are easy to use.

Table **10** lists the maximum NDLs for any series of dives (up to 3) with 60 *min* SIs between dives. Divers need make a deep stop at half the maximum bottom pressure for 1 *min*, plus a shallow safety stop in the 15 *fsw* zone for 2 *min*. Descent rate is 60 *fsw/min*, and ascent rate is 30 *fsw/min*. The NDLs are listed for maximum risk after 3 repetitive dives to the (same) depth indicated, or to a lesser depth.

Table 10. RGBM repetitive risks for air dives.

	β 5.14%	β 3.29%	β 1.37%	
Depth (*fsw*)	**Maximum Time (*min*)**	**Maximum Time (*min*)**	**Maximum Time (*min*)**	
100	24	20	14	deep stop 60/1 shallow stop 15/2

(Table 10) contd.....

	β 5.14%	β 3.29%	β 1.37%	
Depth (*fsw*)	**Maximum Time (*min*)**	**Maximum Time (*min*)**	**Maximum Time (*min*)**	
80	38	32	24	deep stop 50/1 shallow stop 15/2
60	50	42	32	deep stop 40/1 shallow stop 15/2
40	130	120	100	deep stop 30/1 shallow stop 15/2

Tables like these are of interest to Puerto Rican diving fishermen, and fishing sport divers. NAUI uses a variant, detailed next, for training. Technical Training Agencies also employ mixed gas tables for decompression diving, as well as dive planning software, all based on the RGBM algorithm. Some risk estimates of profiles in these RGBM Technical Tables also follow.

NAUI AIR AND NITROX RECREATIONAL TABLES (SEA LEVEL - 10,000 *FT*)

For comparison, consider similar RGBM Tables employed by NAUI for air and nitrox diver training, sea level up to 10,000 ft. They are basically the same as the Puerto Rican seafood diver tables above, except that successive dives must always be shallower than the previous. Descent and ascent rates are 75 *fsw/min* and 30 *fsw/min*, and SIs are 60 *min*. At sea level to 2,000 *ft* elevation, three dives may be made in a day on air or nitrox. At elevations above 2,000 ft, only two dives are sanctioned. There are 9 RGBM Tables in all, 3 for air, 3 for EAN32, and 3 for EAN36, ranging in altitude, 0 - 2,000 ft, 2,000 - 6,000 ft, and 9,000 - 10,000 ft. In Tables **11a** through **11c**, risks are tabulated at the end of the 3 or 2 dive sequence, for just 3 Tables (air at 6,000 - 10,000 *ft*, EAN32 at 2,000 - 6,000 ft, and EAN36 at 0- 2,000 ft). Risks decrease at any elevation as the oxygen fraction increases, while elevation increases risk for any mixture of nitrogen and oxygen. Moving from left to right (first dive through last permitted dive) successive decrements in permissible depths are seen. Safety stops at half the bottom depth are required for 1 - 2 *min*, plus a shallow stop in the 15 *fsw* zone for 2 *min*. The shallow stop mostly serves to control ascent speed. Maximum risk is seen in the air tables at 10,000 *ft* elevation, and minimum risk in the EAN36 tables at sea level.

Table 11a. NAUI RGBM air tables (6,000 - 10,000 *ft*) maximum risk after dive 2, $\beta = 2.36\%$.

Dive 1		Dive 2	
Depth (fsw)	***Time (min)***	***Depth (fsw)***	***Time (min)***
90	11	60	28
80	15	55	28
70	21	50	40
60	28	45	40
50	40	40	64
40	64	35	64
30	103	30	103

Table 11b. NAUI RGBM EAN32 tables (2,000 - 6,000 *ft*) maximum risk after dive 2, $\beta = 1.65\%$.

Dive 1		Dive 2	
Depth (fsw)	***Time (min)***	***Depth (fsw)***	***Time (min)***
100	20	65	43
90	26	60	57
80	33	55	57
70	43	50	84
60	57	45	84
50	84	40	120
40	120	35	120
30	150	30	150

Table 11c. NAUI RGBM EAN36 tables (0 - 2,000 *ft*) maximum risk after dive 3, $\beta = 1.12\%$.

Dive 1		Dive 2		Dive 3	
Depth (fsw)	***Time (min)***	***Depth (fsw)***	***Time (min)***	***Depth (fsw)***	***Time (min)***
110	31	80	60	50	150

(Table 11c) contd.....

Dive 1		**Dive 2**		**Dive 3**	
Depth (fsw)	***Time (min)***	***Depth (fsw)***	***Time (min)***	***Depth (fsw)***	***Time (min)***
100	35	75	60	50	150
90	46	70	85	50	150
80	60	65	85	50	150
70	85	60	115	50	150
60	115	55	115	50	150
50	150	50	150	50	150

These air and nitrox tables have been backbones in NAUI training regimes. They are simple to use, and easy to teach, avoiding USN Group tags.

HELITROX NONSTOP LIMITS (NDLS)

Helitrox is enriched trimix, that is, the oxygen fraction is above 21% in the breathing mixture. Helitrox is gaining in popularity over nitrox when helium is available for gas mixing. Diving agencies often use helitrox in the beginning sequence of technical diver training. Listed below in Table **12** are nonstop time limits and corresponding risks, β, for exposures at that depth-time. The mixture is helitrox (enriched 26/17 trimix), sometimes called triox.

Table 12. Helitrox NDLs and risk.

Depth ***d (fsw)***	**Time** t_n ***(min)***	**Tisk** β
70	35	1.4%
80	25	1.4%
90	20	1.4%
100	15	1.4%
110	10	1.5%
120	8	1.5%
130	6	1.4%
140	5	1.5%
150	4	1.6%

These NDL triox risks track closely with NDL risks for air and nitrox.

COMPARATIVE HELIUM AND NITROGEN STAGING AND RISK

Consider a deep trimix dive with multiple switches on the way up. This is a risky technical dive, performed by seasoned professionals. Table **13** contrasts stop times for two gas choices at the 100 *fsw* switch. The dive is a short 10 *min* at 400 *fsw* on 10/65/25 trimix, with switches at 235 *fsw*, 100 *fsw*, and 30 *fsw*. Descent and ascent rates are 75 *fsw/min* and 25 *fsw/min*. Obviously, there are many other choices for switch depths, mixtures, and strategies. In this comparison, the oxygen fractions were the same in all mixes, at all switches. Differences between a nitrogen or a helium based decompression strategy, even for this short exposure, are nominal. Such usually is the case when oxygen fraction is held constant in helium or nitrogen mixes at the switch.

Comparative profile reports suggest that riding helium to the 70 *fsw* level with a switch to EAN50 is good strategy, one that couples the benefits of well being on helium with minimal decompression time and stress following isobaric switch to nitrogen. Shallower switches to enriched air also work, with only a nominal increase in overall decompression time, but with deeper switches off helium to nitrox a source of isobaric counterdiffusion (ICD) issues that might best be avoided. Note the risk, β, for the helium strategy, 40/20/40 trimix at 100 *fsw*, is slightly safer than the nitrogen strategy, EAN40 at 100 *fsw*, but in either case, the risk is high.

Table 13. Comparative helium and nitrogen gas switches.

	β = 6.42%	**β = 6.97%**
Depth (*fsw*)	**Time (*min*)**	**Time (*min*)**
	10/65 trimix	10/65 trimix
400	10.0	10.0
260	1.5	1.5
250	1.0	1.0
240	1.0	1.0
	18/50 trimix	18/50 trimix
230	0.5	0.5
220	0.5	0.5
210	0.5	0.5

(Table 13) contd.....

	$\beta = 6.42\%$	$\beta = 6.97\%$
Depth (*fsw*)	**Time (*min*)**	**Time (*min*)**
200	0.5	0.5
190	1.0	1.0
180	1.5	1.5
170	1.5	1.0
160	1.5	1.5
150	1.5	2.0
140	2.0	1.5
130	2.0	2.5
120	4.0	4.0
110	4.5	4.0
	40/20 trimix	EAN40
100	2.5	2.0
90	2.5	2.0
80	2.5	2.0
70	5.0	4.0
60	6.5	5.5
50	8.0	6.5
40	9.5	7.5
	EAN80	EAN80
30	10.5	10.5
20	14.0	14.0
10	21.0	20.5
	——-	——-
	123.0	116.0

WKPP EXTREME EXPLORATION DIVES

The Woodville Karst Plain Project (WKPP) has reported a number of 300 *fsw* dives with OC and RB systems on trimix for many hours bottom time, and some 8 *hrs* of decompression. Pure oxygen is employed in the 30 *fsw* zone with the help of an underwater habitat. Success- ful regimens systematically roll back the helium fraction and increase the oxygen fraction in roughly the same proportions, thus maintaining nitrogen fractions low and fairly constant. Diving starts in the cave systems of Wakulla Springs in Florida. Table **14** summarizes the ascent and decompression profile. The risk is high, but WKPP professionals continue to attempt and complete such extreme exposures, pushing the exploration envelope. These dives served as calibration points for the RGBM algorithm on whole.

Table 14. WKPP extreme trimix dives surfacing risk, β = 13.67%.

Depth (*fsw*)	**Time** (*min*)	**Trimix** (*fsw*)	**Depth** (*min*)	**Time**	**Trimix**
270	360	11/50	140	5	
260	1		130	6	
250	1		120	7	35/25
240	1	18/40	110	8	
230	2		100	9	
220	2		90	10	
210	2		80	12	
200	3		70	16	50/16
190	3		60	34	
180	3	21/35	50	41	
170	4		40	49	
160	4		30	60	pure O_2
150	5		20	90	

RECORD OC TRIMIX DIVE

Consider risk after an OC dive to 1040 *fsw* on trimix, with matched ICD switches maintaining the relative fraction of nitrogen constant as helium is reduced in the same measure as oxygen is increased. Dives without this rather well known strategy ended in some serious hyperbaric chamber time for treatment of vestibular DCS. Reports hint this dive was attempted, maybe accomplished, but contradictions abound. We merely treat it as academic exercise for risk prediction. One attempt ended in the Phuket hyperbaric chamber, as reported by a hyperbaric specialist and support team. Earlier dives to 540 *fsw* using RGBM schedules are recounted in trade magazines and at Internet sites. Dives like these with deep stops are becoming more common these days, both on OC and CCR systems.

Table **15** roughly summarizes the RGBM profile and ascent protocol. Stops range from 740 *fsw* to 10 *fsw* for times ranging 0.5 *min* to 31.0 *min*. Descent rate is assumed to be 60 *fsw/min*, and ascent rate between stages is assumed to be 30 *fsw/min*. Mixes and switch depths are indicated, as in Table **14**. Stops are made in 10 *fsw* increments all the way to the surface.

Table 15. Trimix dive to 1040 *fsw* and risk surfacing risk, $\beta = 26.13\%$.

Depth (*fsw*)	Time (*min*)	Trimix	Depth (*fsw*)	Time (*min*)	Trimix
1049	0.5	5/67	380	3.0	
750	0.5		370	3.0	
740	0.5		360	3.0	
730	0.5		350	3.0	
720	0.5		340	3.5	
710	0.5		330	3.5	
700	0.5		320	3.5	
690	0.5		310	3.5	
680	0.5		300	3.5	
670	0.5		290	2.5	14/56
660	0.5		280	3.0	
650	0.5		270	3.5	
640	0.5		260	3.5	
630	0.5		250	3.5	
620	0.5		240	3.5	
610	0.5		230	4.0	
600	0.5		220	4.0	
590	0.5		210	5.0	
580	0.5		200	6.0	
570	1.0		190	6.5	
560	1.0		180	6.5	
550	1.0		170	6.5	
540	1.0		160	7.5	
530	1.0		150	9.0	
520	1.5		140	9.5	
510	1.5		130	8.0	27/56

(Table 15) contd.....

Depth (*fsw*)	Time (*min*)	Trimix	Depth (*fsw*)	Time (*min*)	Trimix
500	1.5		120	8.5	
490	1.5		110	9.0	
480	1.5		100	13.0	
470	1.5		90	13.5	
460	1.5		80	14.0	
450	1.5		70	15.5	
440	2.0		60	16.0	
430	2.0		50	17.5	
420	2.5		40	21.0	
410	2.5		30	22.0	EAN80
400	2.5		20	24.5	
390	3.0		10	31.0	pure O_2

The computed risk for this dive is very high, near 30%. Total decompression time is near 415 *min*. Logistics for stage cylinders are beyond formidable, and the risk for deep support divers is also high.

HYDROSPACE EXPLORER EXTREME RB PROFILE

Table **16** is a deep RB dive downloaded off the HydroSpace EXPLORER computer. From a number of corners, reports of 400 *fsw* dives on rebreather systems are becoming commonplace. Consider this one to 444 *fsw* for 15 *min*. Diluent is 10/85 trimix, and pp_{O2} setpoint is 1.1 *atm*. From a decompression standpoint, rebreather systems are the quickest and most efficient systems for underwater activities. The higher the pp_{O2}, the shorter the overall decompression time. That advantage, however, needs to be played off against increasing risks of oxygen toxicity as oxygen partial pressures increase, especially above 1.4 *atm*. The higher percentage of oxygen and lower percentage of inert gases in higher pp_{O2} setpoints of closed circuit rebreathers (CCRs) results in reduced risks, simply because gas loadings and bubble couplings are less in magnitude and importance. This shows up in any set of comparative pp_{O2} RB calculations, as well as in OC *versus* RB risk estimates.

Table 16. Extreme RB dive and risk surfacing risk $\beta = 5.79\%$.

Depth (*fsw*)	Time (*min*)	Depth (*fsw*)	Time (*min*)
444	15.	150	2.0
290	0.5	140	2.0
280	0.5	130	2.0
270	0.5	120	2.5
260	0.5	110	3.0
250	0.5	100	3.5
240	0.5	90	4.0
230	1.0	80	4.5
220	1.0	70	5.0
210	1.0	60	7.0
200	1.0	50	7.5
190	1.5	40	8.0
180	1.5	30	12.5
170	1.5	20	14.0
160	1.5	10	15.5

Risk associated with this 444 *fsw* dive is less than a similar dive on trimix to roughly the same depth for a shorter period of time, that is, Table **13**.

USS PERRY DEEP RB WRECK DIVES

A team of divers uncovered the wreck of the USS Perry in approximately 250 *fsw* off Anguar, and explored it for a week on RBs. Diving in extremely hazardous and changing currents, their repetitive decompression profile appears in Table **17**. Profiles and risk for the two dives, separated by 4 *hrs* SI, are nominal, with no accounting of exertion effort in current implied. Diluent is 10/50 trimix, with a pp_{O2} setpoint of 1.3 *atm*.

Table 17. USS Perry RB repetitive decompression dives and risk surfacing risk after dive 1, $\beta = 5.32\%$ surfacing risk after dive 2, $\beta = 5.89\%$.

Depth (*fsw*)	Time (*min*)	Depth (*fsw*)	Time (*min*)
260	40	40	5
170	1	30	6
160	1	20	9
150	1	10	12
140	1	0	270
130	1	210	20
120	1	90	1
100	2	80	1
90	2	70	1
80	2	60	1
70	3	50	2
60	3	40	2
50	4	30	9

References

1. Adamson A.W., 1976, The Physical Chemistry Of Surfaces, New York: John Wiley And Sons Incorporated.
2. Batchelor G.K., 1953, Theory Of Homogeneous Turbulence, New York: Cambridge University Press.
3. Bateman J.B. and Lang J., 1945, Formation And Growth Of Bubbles In Aqueous Solutions, Canad. J. Res. E23, 22-31.
4. Bennett P.B., Wienke B.R., and Mitchell S., 2008, Decompression And The Deep Stop Workshop Proceedings, UHMS/NAVSEA: Salt Lake City.
5. Berghage T.E. and Durman D., 1980, US Navy Air Recompression Schedule Risk Analysis, Nav. Med. Res. Bull. 1, 1-22.
6. Bevington P. and Robinson D.C., 2003, Data Reduction And Error Analysis For The Physical Sciences, New York: McGraw-Hill Book Company.
7. Blank M., 1979, Monolayer Permeability, Prog. Surface Mem. Science, 13, 87-139.
8. Bowker A.H. and Lieberman G.J., 1964, Engineering Statistics, Engelwood Cliffs: Prentice- Hall.
9. Boycott A.E., Damant G.C.C., and Haldane J.S., 1908, The Prevention Of Compressed Air Illness, J. Hyg. 8, 342-443.
10. Bradley M.E., 1984, Commercial Diving Fatalities, Aviat. Space Environ. Med. 55, 721-724.
11. Brennen C.E., 1995, Cavitation And Bubble Dynamics, Oxford: Oxford University Press.
12. Brubakk A.O. and Neuman T.S., 2003, Physiology And Medicine Of Diving (5th Edition) Edinburgh: W.B. Saunders.
13. Brubakk A.O., Arntzen A.J., Wienke B.R. and Koteng S., 2003, Decompression Profile And Bubble Formation After Dives With Surface Decompression: Experimental Support For A Dual Phase Model Of Decompression, Undersea Hyper. Med. 30, 181-193.
14. Buckles R.G., 1968, The Physics Of Bubble Formation And Growth, Aerospace Med. 39, 1062-1069.
15. Buhlmann A.A., 1984, Decompression/Decompression Sickness, Berlin: Springer Verlag.
16. Buhlmann A.A., 1966, Saturation And Desaturation With N_2 And He At 4 Atmospheres, J. Appl. Physiol. 23, 458-462.
17. Butler F.K. and Southerland D.K., 2001, The US Navy Decompression Computerr,

Undersea Hyper. Med. 28, 213-228. Circuit Scuba Divers, Undersea Biomed. Res. 13, 193-223.

18. Campbell J., 1968, The Tribonucleation Of Bubbles, Brit J. Appl. Phys. 1, 1083-1088.
19. Conkin J. and Van Liew H.D., 1991, Failure Of The Straight Line Boundary Between Safe And Unsafe Decompressions When Extraoplated To The Hypobaric Regime, Under- sea Biomed. Res. 18, 16.
20. Des Granges M., 1957, Repetitive Diving Decompression Tables, USN Experimental Div- ing Unit Report, NEDU 6-57, Washington DC. Decompression Sickness, J. Appl. Physiol. 94, 2145-2150.
21. Duffner G.J., Synder J.F., and Smith L.L., 1959, Adaptation Of Helium-Oxygen To Mixed Gas Scuba, USN Experimental Diving Unit Report, NEDU 3-59, Washington, DC
22. Dunford R., Wacholz C., and Huggins K., 1993, DCS Risk And Doppler Bubbles In Sport Divers, Undersea Hyper. Med. 20, 80.
23. Dwyer J.V., 1956, Calculation Of Repetitive Diving Decompression Tables, USN Experimental Diving Unit Report, NEDU 1-57, Washington DC.
24. Eckenhoff R.G., Olstad C.E., Parker S.F. and Bondi K.R., 1986, Direct Ascent From Shallow Air Saturation Exposures, Undersea Biomed. Res. 13, 305-316.
25. Eckenhoff R.G. and Vann R.D., 1985, Air And Nitrox Saturation Decompression, Undersea Hyper. Med. 12, 41-52.
26. Eckenhoff R.G., 1985, Doppler Bubble Detection, Undersea Biomed. Res. 12, 485-489.
27. Egi S.M. and Brubakk A.O., 1995, Diving At Altitude: Review Of Decompression Strategies, Undersea Hyper. Med. 22, 281-300. Epstein P.S. and Plesset M.S., 1950, On The Stability Of Gas Bubbles In Liquid-Gas Solutions, J. Chem. Phys. 18, 1505-1509.
28. Evans A. and Walder D.N., 1969, Significance Of Gas Macronuclei In The Aetiology Of Decompression Sickness, Nature 222, 251-252.
29. Farm F.P., Hayashi E.M., and Beckman E.L., 1986, Diving And Decompression Sickness Treatment Practices Among Hawaii's Diving Fisherman, University of Hawaii Sea Grant Report, UNIHI-SEAGRANT-TP-86-01, Honolulu.
30. Frenkel J., 1946, Kinetic Theory Of Liquids, New York: Oxford University Press.
31. Gernhardt M.L., Lambertsen C.J., Miller R.G., and Hopkins E., 1990, Evaluation Of A Theoretical Model Of Tissue Gas Phase Growth And Resolution During Decompression From Air Diving, Undersea Biomed. Res. 17, 95.
32. Goldstein H., 1969, Mechanics, Reading: Addison Wesley Publishing Company.

33. Hamilton R.W., 1989, Tolerating Exposure To High Oxygen Levels, REPEX And Other Methods, Marine Tech. Soc. J. 23, 19-25.
34. Hamilton R.W. and Turner P., 1988, Decompression Techniques Based On Special Gas Mixes For Deep Exploration, Undersea Biomed. Res. 15, 70.
35. Hempleman H.V., 1957, Further Basic Facts On Decompression Sickness, Investigation Into The Decompression Tables, Medical Research Council Report, UPS 168, London.
36. Hempleman H.V., 1952, A New Theoretical Basis For The Calculation Of Decompression Tables, Medical research Council Report, UPS 131, London.
37. Hennessy T.R. and Hempleman H.V., 1977, An Examination Of The Critical Released Gas Concept In Decompression Sickness, Proc. Royal Soc. London B197, 299-313.
38. Hills B.A., 1977, Decompression Sickness, New York: John Wiley And Sons Incorporated.
39. Hills B.A., 1976, Supersaturation By Counterdiffusion And Diffusion Of Gases, J. Appl. Physiol. 43, 56-69.
40. Hills B.A., 1968, Variation In Susceptibility To Decompression Sickness, Int. J. Biometeor. 12, 343-349.
41. Ikels K. G., 1970, Production Of Gas Bubbles In Fluids By Tribonucleation, J. Appl. Physiol. 28, 524-527.
42. Kahaner D., Moler C., and Nash S., 1989, Numerical Methods And Software, Englewood Cliffs: Prentice Hall.
43. Keller H. and Buhlmann A.A., 1965, Deep Diving And Short Decompression By Breathing Mixed Gases, J. Appl. Physiol. 20, 1267.
44. Lambertsen J.L. and Bornmann R.C., 1979, Isobaric Inert Gas Counterdiffusion, Undersea And Hyperbaric Medical Society Publication 54WS(IC)1-11-82, Bethesda.
45. Lambertsen C.J. and Idicula J., 1975, A New Gas Lesion Syndrome In Man Induced By Isobaric Gas Counterdiffusion, J. Appl. Physiol. 49, 1070-1082.
46. Lango T., Morland T., and Brubakk A.O., 1996, Diffusion Coefficients And Solubility Coefficients For Gases In Biological Fluids, Undersea Hyper. Med. 23, 247-242.
47. Lehner C.E., Lin T.F., Taya F., Wienke B.R., Nordheim E.V., Cuddon P.A., and Lanphier E.H., 1994, Acclimatization Reduces The Incidence Of Decompression Sickness: A Sheep Model, Undersea Hyper. Med. 21, 22.
48. Le Messurier D.H. and Hills B.A., 1965, Decompression Sickness: A Study Of Diving Techniques In The Torres Strait, Hvaldradets Skrifter 48, 54-84.
49. Neuman T.S., Hall D.A. and Linaweaver P.G., 1976, Gas Phase Separation During

Decompression In Man, Undersea Biomed. Res. 7, 107-112.

50. Nishi R.Y., 1991, Development Of Surface Decompression Tables For Helium-Oxygen Diving To Depths Of 100 msw, Undersea Biomed. Res. 18, 66-67.
51. Nishi R.Y., 1989, Development Of New Helium-Oxygen Decompression Tables For Depths To 100 msw, Undersea Biomed. Res. 16, 26-27.
52. Parzen E., 1970, Modern Probability Theory And Its Applications, New York: John Wiley And Sons Incorporated.
53. Pilmanis A.A., 1976, Intravenous Gas Emboli In Man After Compressed Air Ocean Diving, Office Of Naval Research Contract Report, N00014-67-A-0269-0026, Washington, DC
54. Powell M.P.and Rogers R.E., 1989, Doppler Ultrasound Monitoring Of Gas Phase Formation And Resolution In Repetitive Diving, Undersea Biomed. Res. 16, 69.
55. Rogers R.E. and Powell M.R., 1989, Controlled Hyperbaric Chamber Tests Of Multiday Repetitive Dives, Undersea Biomed. Res. 16, 68.
56. Schreiner H.R. and Hamilton R.W., 1987, Validation Of Decompression Tables, Undersea And Hyperbaric Medical Society Publication 74 (VAL), Bethesda.
57. Sears F.W., 1969, Thermodynamics, Reading: Addison Wesley Publishing Company.
58. Shapiro A.H., 1958, Dynamics And Thermodynamics Of Compressible Fluid Flow, New York: Ronald.
59. Skripov V.P., 1974, Metastable Liquids, New York: John Wiley And Sons Incorporated.
60. Smith K.H. and Stayton L., 1978, Hyperbaric Decompression By Means Of Bubble Detection, Office Of Naval Research Report, N0001-469-C-0402, Washington DC
61. Spencer M.P., 1976, Decompression Limits For Compressed Air Determined By Ultra- sonically Detected Blood Bubbles, J. Appl. Physiol. 40, 229-235
62. Thalmann E.D., 1989, Testing Of Revised Unlimited Duration Upward Excursions During Helium-Oxygen Saturation Dives, Undersea Biomed. Res. 16, 195-214.
63. Thalmann E.D., 1985, Air Tables Revisited, Undersea Biomed. Res. 12, 54.
64. Van Liew H.D. and Hlastala M.P., 1969, Influence Of Bubble Size And Blood Perfusion On Absorption Of Gas Bubbles In Tissues, Resp. Physiol. 24, 111-121.
65. Van Liew H.D., Bishop B, Walder P.D., and Rahn H., 1975, Bubble Growth And Mechanical Properties Of Tissue In Decompression, Undersea Biomed. Res. 2, 185-194.
66. Vann R.D. and Uguccioni D.M., 2001, DAN Report On Decompression Illness, Diving Fatalities, And Project Dive Exploration, Durham: Divers Alert Network.
67. Vann R.D., Gerth W.A., and Charlton W.A., 1999, Operational Testing Of Air And Nitrox Dive Tables, Undersea Hyper. Med. 26, 48.

68. Vann R.D. and Gerth W.A., 1997, Probabilistic Gas And Bubble Dynamics Models Of Decompression Sickness Occurrence In Air And Nitrogen-Oxygen Diving, Undersea Hyper. Med. 24, 275-292.
69. Vann R.D., Gerth W.A., and Leatherman N.E., 1987, A Likelihood Analysis Of Experiments To Test Altitude Decompression Protocols For SHUTTLE Operations, Aviat. Space Environ. Med. 58, A106-A109.
70. Ward C.A., Balakrishnan A., and Hooper F.C., 1979, On The Thermodynamics Of Nucleation In Weak Gas Solutions, J. Basic Eng. 92, 695-704.
71. Weathersby P.K., Hart B.I., and Flynn E.T., 1987, Role Of Oxygen In The Production Of Human Decompression Sickness, J. Appl. Physiol. 63, 2380-2387.
72. Weathersby P.K., Survanshi S. and Homer L.D., 1985, Statistically Based Decompression Tables: Analysis Of Standard Air Dives, 1950-1970, Naval Medical Research Institute Report, NMRI 85-16, Bethesda.
73. Weathersby P.K., Homer L.D., and Flynn E.T., 1984, On The Likelihood Of Decompression Sickness, J. Appl. Physiol. 57, 815-825.
74. Wienke B.R., 2015, Deep Stop Model Correlations, J. Bioeng. Biomed. Sci. 5, 12-18.
75. Wienke B.R., 2015, Science Of Diving: Concepts And Applications, Boca Raton: CRC Press (Taylor And Francis Group).
76. Wienke B.R., 2010, Computer Validation And Statistical Correlations Of A Modern Decompression Diving Algorithm, Comp. Biol. Med. 40, 252-260.
77. Wienke B.R. and O'Leary T.R., 2008, Statistical Correlations And Risk Analysis Techniques For A Diving Dual Phase Bubble Model And Data Bank Using Massively Parallel Supercomputers, Comp. Biol. Med. 38, 583-600.
78. Wienke B.R., 1992, Numerical Phase Algorithm For Decompression Computers And Application, Comp. Biol. Med. 22, 389-406.
79. Wienke B.R., 1991, Bubble Number Saturation Curve And Asymptotics Of Hypobaric And Hyperbaric Exposures, Int. J. Biomed. Comp. 29, 215-225.
80. Wienke B.R., 1990, Reduced Gradient Bubble Model, Int. J. Biomed. Comp. 26, 237-256.
81. Wienke B.R., 1989, Equivalent Multitissue And Thermodynamic Decompression Algorithms, Int. J. BioMed. Comp. 24, 227-245.
82. Wienke B.R., 1989, Tissue Gas Exchange Models And Decompression Computations: A Review, Undersea Biomed. Res. 16, 53-89.
83. Wienke B.R., 1987, Computational Decompression Models, Int. J. BioMed. Comp. 21, 205-221.
84. Wienke B.R., 1986, Phenomenolgical Models For Nitrogen Transport In Tissues, Il Nuovo Cimento 8D, 417-435.

85. Workman R.D., 1965, Calculation Of Decompression Schedules For Nitrogen-Oxygen And Helium-Oxygen Dives, USN Experimental Diving Unit Report, NEDU 6-65, Washington DC
86. Yang W.J., 1971, Dynamics Of Gas Bubbles In Whole Blood And Plasma, J. Biomech. 4, 119-125.
87. Young F.R., 1989, Cavitation, New York: McGraw-Hill Book Company.
88. Young C. and D'Aoust B.G., 1981, Factors Determining Temporal Patterns Of Isobaric Supersaturation, J. Appl. Physiol. 51, 852-857.
89. Yount D.E. and Hoffman DC, 1986, On The Use Of A Bubble Formation Model To Calculate Diving Tables, Aviat. Space Environ. Med. 57, 149-156.
90. Yount D.E., 1982, On The Evolution, Generation, And Regeneration Of Gas Cavitation Nuclei, J. Acoust. Soc. Am. 71, 1473-1481.
91. Yount D.E., 1979, Skins Of Varying Permeability: A Stabilization Mechanism For Gas Cavitation Nuclei, J. Acoust. Soc. Am. 65, 1431-1439.
92. Balestra C.,2012, Dive Computer Use In Recreational Diving: Insights From The DAN- DSL Database, Proceedings Validation Of Dive Computers Workshop, EUBS: Gdansk.
93. Gilliam B., Maier V., and Crea J., 1995, Deep Diving, San Diego: Watersports Publish- ing.
94. Maronni A, Bennett PB, Cronje FJ, et al., 2004, Effect Of Varying Deep Stop Times And Shallow Stop Times On Precordial Bubble Scores After Dives To 35 msw, Undersea Hyper. Med. 31, 233-243.
95. Gerth W.A., Doolette D.J., and Gault K.A., 2008, Deep Stops And Their Efficacy In Decompression: US Navy Research, Decompression And The Deep Stop Workshop Pro- ceedings, UHMS/NAVSEA: Salt Lake City.
96. Blatteau J.E., Hugon M., and Gardette B., 2008, Deep Stops During Decompression From 50 To 100 msw Didn't Reduce Bubble Formation In Man, Decompression And The Deep Stop Workshop Proceedings, UHMS/NAVSEA: Salt Lake City.

SENTINEL REFERENCES

Some important references in the above list include:

1. Bennett P.B., Wienke B.R., and Mitchell S., 2008, Decompression And The Deep Stop Workshop Proceedings, UHMS/NAVSEA: Salt Lake City.
2. Buckles R.G., 1968, The Physics Of Bubble Formation And Growth, Aerospace Med. 39, 1062-1069.
3. Hennessy T.R. and Hemplemen H.V., 1977, An Examination Of The Critical

Released Gas Concept In Decompression Sickness, Proc. Royal Soc. London B197, 299-313.

4. Hills B.A., 1977, Decompression Sickness, New York: John Wiley and Sons Inc.
5. LeMessurier D.H. and Hills B.A., 1965, Decompression Sickness: A Study Of Diving Techniques In The Torres Strait, Hvaldradets 48, 54-84.
6. Ward C.A., Balakrishnan A., and Hooper F.C., 1979, On The Thermodynamics Of Nucleation In Weak Gas Solutions, J. Basic Eng. 92, 695-704.
7. Weathersby P.K., Homer L.D., and Flynn E.T., 1984, On the Likelihood Of Decompres- sion Sickness, L. Appl. Physiol. 57, 815-825.
8. Wienke B.R., 2015, Science Of Diving: Concepts And Applications, Boca Raton: CRC Press (Taylor and Francis Group).
9. Wienke B.R., 2010, Computer Validation And Statistical Correlations Of A Modern Decompression Diving Algorithm, Comp. Biol. Med. 40, 252-260.
10. Wienke B.R., 2009, Diving Decompression Models And Bubble Metrics: Modern Computer Syntheses, Comp. Biol. Med. 39, 309-311.
11. Young F.R., 1989, Cavitation, New York: McGraw-Hill Book Company.
12. Yount D.E., 1982, On The Evolution, Generation, And Regeneration Of Gas Cavitation Nuclei, J. Acoust. Soc. Am. 71, 1473-1481.

BIOSKETCH

Bruce Wienke is a Program Manager in the Nuclear Weapons Technology/ Simulation And Computing Office at the Los Alamos National Laboratory (LANL), with interests in hydrodynamics, applied mathematics, particle and nuclear physics, numerical methods, parallel computing, thermonuclear burn, decompression phenomenology and models, radiation and gas transport, bubble dynamics, and phase mechanics, publishing some 240+ technical and research papers in related Journals. He contributes to underwater symposia, educational publications, technical periodicals and decompression workshops, authoring ten monographs *(Computational Decompression Theory And Applications, Science Of Diving, Hyperbaric Physics With Bubble Mechanics And Decompression Theory, Reduced Gradient Bubble Model In Depth, Technical Diving In Depth, Physics, Physiology And Decompression Theory For The Technical And Commercial Diver, High Altitude Diving, Basic Diving Physics And Applications, Diving Above Sea Level, Basic Decompression Theory And Applications).* Diving experience includes the Caribbean, South Pacific, Asia, inland and coastal United States, Hawaii, and polar Arctic and Antarctic on OC and RB systems in technical, scientific, military, and research exercises. He functions on the LANL Nuclear Emergency Strategy Team (NEST), above and below water, heading up the C & C Dive Team.

Wienke is Workshop Director/Trainer with the National Association Of Underwater Instructors (NAUI), and served on the Board Of Directors (Vice Chairman for Technical Diving, Technical and Decompression Review Board Member). Wintertime he hobbies ski racing, coaching, and teaching. As a Racing Coach and Instructor, he is certified United States Ski Coaches Association (USSCA) and Professional Ski Instructors of America (PSIA), and competes in the United States Ski Association (USSA) Masters Series, holding a 8 NASTAR racing handicap while winning NASTAR National Championships in his age class in 2002 and 2011. He quarterbacked the Northern Michigan Wildcats to an NCAA-II National Championship in 1963, garnering All American Honors. Other pastimes include tennis, windsurfing, and mountain biking.

Wienke received a BS in physics and mathematics from Northern Michigan University, MS in nuclear physics from Marquette University, and PhD in particle physics from Northwestern University. He belongs to the American Physical Society (APS), American Nuclear Society (ANS), Society Of Industrial And Applied Mathematics (SIAM), South Pacific Underwater Medical Society (SPUMS), Undersea And Hyperbaric Medical Society (UHMS), and American Academy Of Underwater Sciences (AAUS). He is a Fellow of the American Physical Society, and a Technical Committee Member of the American Nuclear Society.

Wienke advises on decompression algorithms across exploration, recreational, technical, commercial, scientific, and research sectors, and developed the reduced gradient bubble model (RGBM), a dual phase approach to staging diver ascents over an extended range of diving applications (altitude, nonstop, decompression, multiday, repetitive, multilevel, mixed gas, and saturation). Many modern dive computers incorporate the modified and iterative RGBM into staging regimens for OC and RB mixed gas diving. In lock step, computer software and platforms offer RGBM for technical dive planning and profile analysis. A number of Training Agencies employ RGBM tables, software, and meters for hands on training and education within their Standards and Procedures. Wienke also brought the LANL Data Bank online for model and profile analysis and risk assessment. The LANL Data Bank is similar to the DAN PDE Data Bank but focuses on tec diving data. The LANL Data Bank presently (as of 2/16) contains 3023 mixed gas, OC and RB, decompression profiles with 23 cases of DCS. The Data Bank has been instrumental in RGBM success worldwide and a valued component of C&C dive planning over the past 12 years.

Wienke serves as Reviewer Editor for the *Journal of Computational Physics, Physical Review, Transport Theory And Statistical Mechanics, Applied Physics, Nuclear Science And Engineering, Nuclear Fusion, Journal Of Quantitative Spectroscopy And Radiation Transport, Nuovo Cimento, and Journal of Applied Physics.* He is also an Associate Editor for the *International Journal Of Aquatic Research And Education*, a Contributing Editor of Sources, the NAUI training periodical, and a Contributing Editor of *Advanced Diver Magazine*, a trade publication for technical diving.

SUBJECT INDEX

B.R. Wienke (Ed.)

C

E

F

M

N

O

P

T

U

V

W

Y

Z

Made in United States
Orlando, FL
04 November 2026

90199045R00100